CRYSTALS IN GLASS

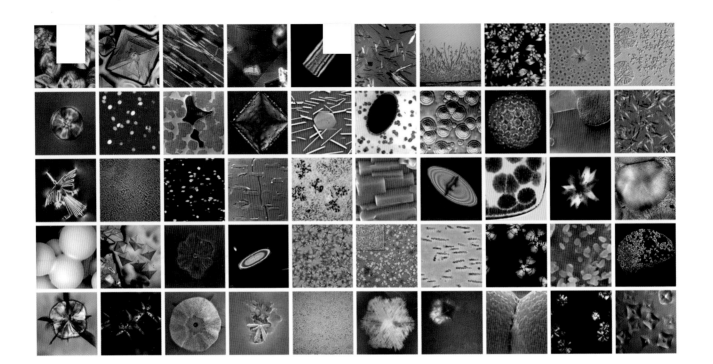

Edgar D. Zanotto

CRYSTALS IN GLASS

A Hidden Beauty

Cover Design: John Wiley & Sons, Inc.
Cover Photograph: © *Edgar D. Zanotto*

Published by John Wiley & Sons, Inc., Hoboken, New Jersey.
Published simultaneously in Canada.

For general information on our other products and services or for technical support, please contact our Customer Care Department within the United States at (800) 762-2974, outside the United States at (317) 572-3993 or fax (317) 572-4002.

Wiley also publishes its books in a variety of electronic formats. Some content that appears in print may not be available in electronic formats. For more information about Wiley products, visit our web site at www.wiley.com.

Library of Congress Cataloging-in-Publication Data:

Zanotto, E. D. (Edgar Dutra)
  Crystals in glass : a hidden beauty / by Edgar D. Zanotto.
    pages cm
  Includes bibliographical references and index.
  ISBN 978-1-118-52143-4 (cloth : alk. paper) 1. Glass-ceramics. 2. Silicate crystals.
3. Nucleation. 4. Crystallization. I. Title.
  TP862.Z36 2013
  660'.284298–dc23
                                                    2012045177

Printed in Singapore.

10 9 8 7 6 5 4 3 2 1

CONTENTS

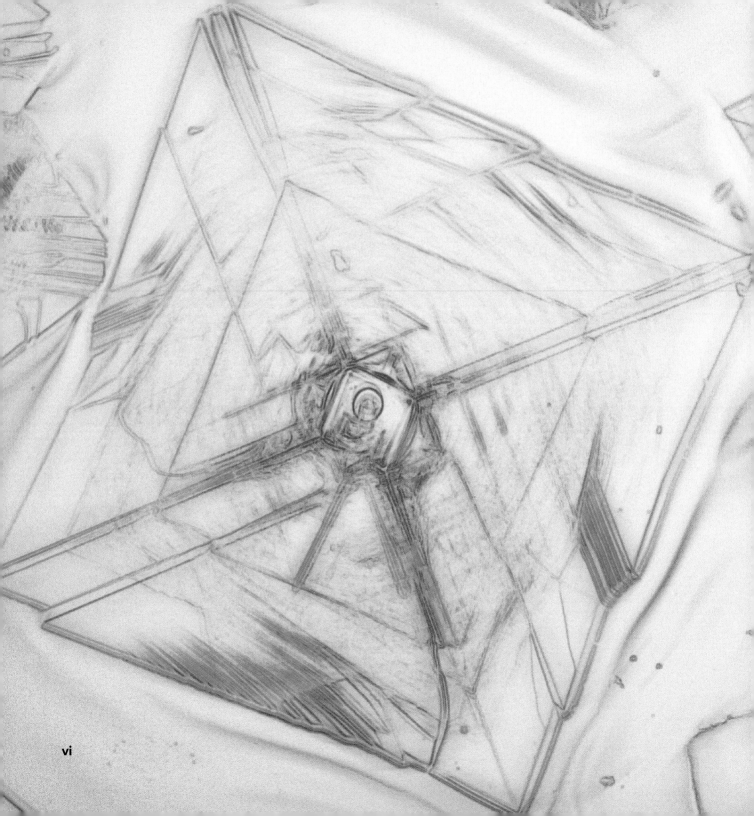

FOREWORD

Many years ago, when I first had the opportunity to see his mind at work, Edgar Zanotto offered me a living preview of this book. By then, in the mid-1990s, he was already famous as the father of glass science in Brazil, and within my department his entrepreneurship was viewed as a model. His first advisor, Aldo Craievich, was a highly respected former member of our faculty, and their work was often cited as a fine example of physical insight. Nonetheless, Zanotto and I had only known each other superficially until our jobs at the State of São Paulo funding agency (FAPESP) brought us together.

After that, besides sharing an office, we frequently rode the same car between São Carlos and São Paulo. In the office, in restaurants, or on the highway, he gave accounts of ideas that were being cast into papers or patents and explained the hows and whys of glassy materials. From devitrification in glass bottles to the courtyard effects, our conversations covered countless aspects of glass science and technology. They were, nonetheless, almost always focused on ongoing projects or past experiences. Inspirations were rarely discussed, let alone dreams. And so it was that the concept of a book was never brought up, although all of Edgar's friends knew that even a partial sum of his achievements would add up to an attractive volume.

The book is now ready, much more radiant than one could have imagined in those days. Part of the glow comes from the micrographs chosen by the author to illustrate his story. Each bit of reasoning in the book is supported by a picture, but the micrographs are more than simple illustrations: they constitute the conducting thread that drives our imagination from the first to the last page, from the *Drosophila melanogaster* of glass crystallization to bubble nucleation in bioactive glass-ceramics. This thread takes us on a ride through roads lined with murals, as it were, covering the elements of the science developed in Zanotto's LaMaV. The pictures spur the reader's imagination, and a few of them, such as the eloquent lessons on competition depicted on pages 39 and 85, are lectures in nutshells.

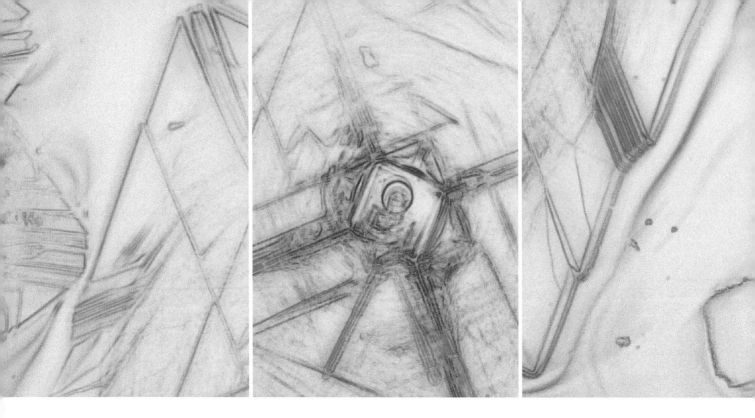

Edgar Zanotto is one of the world's top experts on crystallization in glasses, and his experience is imprinted on every page. The historical account in the Introduction and the motto at the façade of the LaMaV, photographed on page xiii, are likely to motivate young researchers, but the author's skill is most clearly revealed in the selection of micrographs and in the accompanying texts. The latter are gems on their own. Aware that a book written in pictorial language would be poorly served by prolix paragraphs, Zanotto has crafted sentences with the proper dose of information to heighten our scientific appetite.

The result is not a textbook, a tutorial, or a science treatise. Nor is it a coffee-table book, meant to please the eyes and sooth the spirit. To be sure, the visual component of the book is outstanding, so beautiful are the pictures between its covers. Yet, Zanotto's brainchild is above all a source of inspiration. In each picture, we find challenges that have compelled humanity since the dawn of civilization: "Crack my code, relate me to other elements of your universe, and take control of my dynamics," the crystallites seem to be crying out. Each plate invites us to read the facing text and to meditate about the big and the small, about symmetries and broken symmetries, and about time and space. Most readers will therefore find it difficult to skip rapidly through the pages.

Different readers have different styles. Linear sequencing displeases many of us. We enjoy moving back and forth between pages to let our attention be caught at haphazard. To all adepts of this unorthodox reading procedure, "Crystals in Glasses: a Hidden Beauty" comes close to perfection. Although loosely divided into five sections, the pictures and the descriptions are independent and can be examined in any order.

This method will be called kaleidoscopic by partisans of orderly reading. Kaleidoscopes, however, are but disappointing combinations of symmetry with randomness, perhaps designed to let children know that chance and shattering are parts of life. Here, Edgar Zanotto reinvents the kaleidoscope. With 50 examples, he shows that little pieces of glass hide not only beauty but also lessons and challenges that may lift our spirit and consequently strengthen our belief in the meaning of life.

Luiz Nunes Oliveira
Professor of Physics
Institute of Physics of São Carlos (IFSC)
University of São Paulo (USP), Brazil

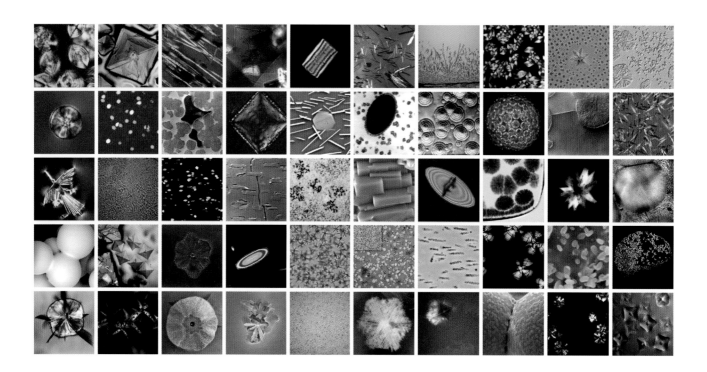

INTRODUCTION

36 Years of Research and Discoveries about Glass Crystallization

With this technically biased introduction, I hope to make the scientist readership aware of and persuade it to browse through some of our discoveries on glass crystallization published over the past 3.5 decades, while cordially inviting the layman to skip the text below and go directly to the photographic exhibition.

The Vitreous Materials Lab (LaMaV), Department of Materials Engineering (DEMa), at the Federal University of São Carlos (UFSCar) was founded on December 15, 1976, when I, then a young materials engineer, was hired by the UFSCar. To celebrate LaMaV's 36th anniversary, I briefly review the most significant scientific work carried out by our team and collaborators and present a selected collection of extraordinary micrographs (photographs taken under an optical or electron microscope) that not only reveal some intricate secrets about glass crystallization but also display the hidden beauty of microscopic crystals!

LaMaV's research work focuses mainly on fundamental studies of glass crystallization kinetics and mechanisms, which are key to the successful development of *glass-ceramics* (CGs). In this field, LaMaV's contributions are one of the world's outstanding.

CGs are polycrystalline materials produced by the controlled crystallization of certain glasses that possess unusual properties. Natural CGs, such as some types of obsidian, have "always" existed. But in 1953, Stanley D. Stookey, of Corning Glass Works, USA, made a serendipitous discovery when a furnace containing a piece of lithium disilicate glass with precipitated silver particles (meant to form a permanent photographic image) was accidentally overheated to about 900 °C, although

Stookey actually intended to anneal the glass at 600 °C. Instead of a pool of melted glass, however, the astonished Stookey observed a white material whose shape remained unchanged. Then he accidentally dropped the piece on the floor, but it did not shatter, contrary to what one would normally expect of a piece of glass! Stookey had unwittingly created the first synthetic CG, called Fotoceram. In their famous book on CGs, Wolfram Höland and George Beall state *"knowledge of the literature, good observation skills, and deductive reasoning were clearly evident in allowing the chance events to bear fruit."*

CGs are produced by the controlled crystallization of certain glasses—generally induced by nucleating additives—in contrast to spontaneous surface crystallization, which is normally undesirable in glass manufacturing. CGs always contain a residual glassy phase and one or more embedded crystalline phases, with widely varying crystallinity ranging from 0.5% to 99.5%, most frequently 30–70%. Controlled crystallization yields an array of materials with very interesting, sometimes unusual, combinations of properties. The main advantages of CGs are that, in principle, they can be mass-produced by any glass-forming technique, their nano- or microstructure can be designed for a given application, they have zero or very low porosity, and the desired properties can be combined, including very low thermal expansion coefficient with transparency in the visible wavelength range, for instance, for cooking ware, or very high strength and toughness with translucency, biocompatibility,

chemical durability, and relatively low hardness, for instance, for dental applications. For all these reasons, CGs have found numerous applications, from domestic products to high-tech areas, such as large telescopic mirrors, substrates of hard disks, and artificial bones and teeth.

The main results of LaMaV's research are summarized in the following paragraphs. LaMaV's research on phase transformation kinetics started in 1977, with E. D. Zanotto's MSc dissertation and continued during his PhD studies. Due to the strong controversy in the international literature of the late 1970s, he and his supervisors (Aldo Craievich and the late Peter James) performed detailed studies to establish the effects of *amorphous phase separation* (APS) on the kinetics of crystal nucleation and growth in glasses. Their work demonstrated unequivocally that the compositional shift caused by APS is the main factor responsible for the enhancement of nucleation and growth rates, with the interfaces between the amorphous phases playing a very minor role. They have published seven original papers on this specific topic.

Next, with the invaluable help of the late Mike Weinberg, the LaMaV team spent many years testing the validity of the Classical Nucleation Theory (*CNT*) in different ways with various glasses (a task that has not yet been concluded!). In their first study, they used a constant value of surface energy, σ_∞, and assumed that the activation enthalpy for atomic jumps at the liquid/crystal nucleus interface (ΔG_η) was similar to that of viscous flow (ΔG_d), a procedure also adopted by

other authors. A large discrepancy was found between the experimental nucleation rates and the theoretical values, although the temperature dependence was well described by theory. In a subsequent study, an original and more rigorous approach was proposed and tested using the transient times for nucleation instead of the viscosity to calculate ΔG_{d}. Again, significant discrepancies were found in the magnitude and temperature dependence of the nucleation rates of four glasses. Finally, the most severe assumption of CNT was tested, that is, the postulation that the interfacial energy σ does not depend on nucleus size: $\sigma = \sigma_{\infty}$. In this case, the Tolman expression for $\sigma(r)$ was incorporated into CNT. An impressive agreement between theory and experiment was found. However, $\sigma(r)$ could not be determined independently in all the tests and had to be fitted by forcing the temperature of the experimental and theoretical maxima to coincide. For a definitive test, an independent determination of $\sigma(r)$ will be necessary. Several scientists around the world, including LaMaV's closest collaborators, Juern Schmelzer and Vlad Fokin, are still pursuing this objective. The above-described research and some related studies are original and are described in 13 papers.

The possibility of *metastable phase* precipitation in the early stages of crystallization is of fundamental importance because, in principle, if confirmed, it could explain the frequent discrepancy between CNT predictions and experimental nucleation rates. With the invaluable help of Mike Weinberg, Pierre Lucas and a former student, Paulo Soares, LaMaV's research group established that metastable phases do in fact appear in some glasses, but do not explain the discrepancies in CNT. The group has published eight papers on this particular phenomenon.

The LaMaV group then conducted the first rigorous test of the General Theory of Transformation Kinetics (the *JMAK theory*), with no adjustable parameters, for both homogeneous and heterogeneous nucleation. They demonstrated that, provided proper precautions are taken in the determination of the nucleation and growth rates, the JMAK theory offers an exceptionally accurate description of the time evolution of overall crystallization of stoichiometric glasses. The group has published four papers on this particular subject.

Next, a series of 10 papers discussed the following question: *Why is the thermodynamically unfavorable mechanism—homogeneous nucleation—observed in some glasses?* One of the team's first studies demonstrated that a simple rule of thumb can be used to distinguish the nucleation mechanism in several stoichiometric glass-forming systems. For compositions having glass-transition temperatures ($T_g/T_f < 0.6$), the temperatures of maximum nucleation rates, T_{max}, are higher than T_g. These systems show homogeneous (internal) nucleation on a laboratory timescale. On the other hand, for the majority of glasses, the typical values of reduced T_g are high ($T_g/T_f > 0.6$), the calculated (by CNT) values of T_{max} are significantly lower than T_g, and only heterogeneous nucleation is observed. No exceptions to this remarkable trend have been reported so far. In a second study, the LaMaV team demonstrated that the failure to detect homogeneous nucleation in the other family of glasses (for which $T_{max} < T_g$) is due to one or both of the following causes: excessively low nucleation or growth rates, or long induction times for nucleation in the temperature ranges in which the predicted steady-state homogeneous nucleation rates are expected to be significant.

Later, in collaboration with visiting scientists Eberhard Mueller and Klaus Heide, the LaMaV team showed that the mass densities of glasses that nucleate homogeneously are similar to those of their crystal phases, while those of glasses which only nucleate heterogeneously are quite different from the densities of the equivalent crystals. In the final part of this research, they unambiguously demonstrated that glasses presenting homogeneous nucleation have both cation and anion arrangements that are very similar to their corresponding crystal phases, while the structures of the glass and crystal phases of compositions that only nucleate heterogeneously are quite distinct. Several other revealing papers have been published about the relationships between the molecular *structure* and *nucleation* mechanisms, with

the invaluable participation of Joe Zwanziger, Valmor Mastelaro, Jose Schneider, and other researchers. Some articles have been published on this issue.

Two papers discussed *stereological errors* associated with experimental measurements of nucleation and overall crystallization rates in glasses, which are often neglected in the literature. In some situations, such errors can be quite significant (>50%) and must be taken into account for a proper analysis of crystallization kinetics.

Next, LaMaV's research group, with the invaluable help of former postdoc Nora Mora and visiting scientists Ralf Mueller and Vlad Fokin, focused on the establishment of the mechanisms and kinetics of crystal nucleation and growth on glass *surfaces*. A few papers published by this team triggered international interest in the subject. In those articles, they concluded that, in most cases, surface nucleation is immeasurably fast and is triggered by a certain number of impurity sites, which depend on the degree of surface perfection and cleanliness. The high surface nucleation rates are due to the low interfacial energy between contaminants and nuclei existing on the surface. The resulting papers were of fundamental importance for the development of commercial sintered CGs, which will be described later in this review. The group has published 14 papers on this subject.

Based on their accumulated experience with surface crystallization, the group started to develop systematic research into glass *sintering* with *concurrent crystallization*. This is an issue of keen interest because all sorts of complex-shaped monolithic glasses and CGs can be produced by controlled sintering of powdered glasses. The key problem here is to privilege viscous flow sintering before surface crystallization takes over and causes densification. With the energy and scientific focus of former postdocs Miguel Prado, Catia Fredericci, Vivi Soares, and Anne Barbosa, PhD student Rapha Reis,

and a few other collaborators, they produced quite conclusive and practical results on this subject, which are summarized in 11 papers. Of special interest is the last paper in the series, which proposes and successfully tests a useful sinterability parameter.

LaMaV's research work also included incursions into the complex field of *polymer crystallization*, with five papers published in leading journals. Research studies have also been developed by the LaMaV group on

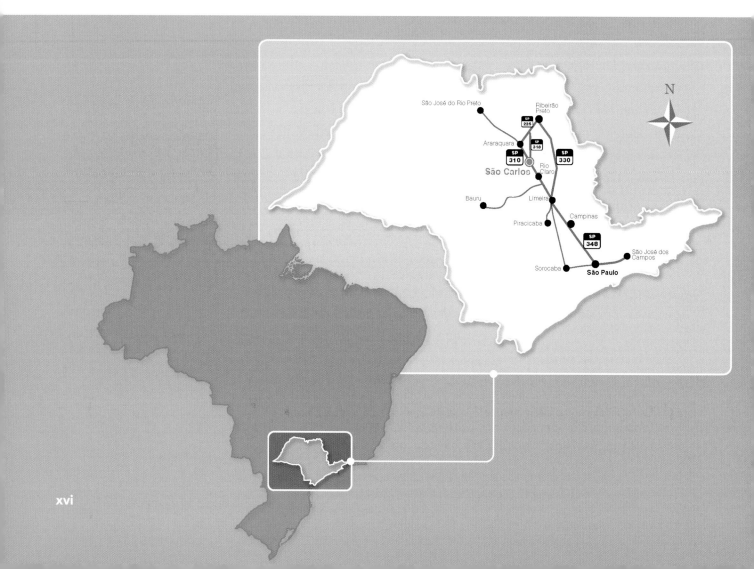

the fundamental issue of *glass-forming ability* of different liquid systems. Several original findings are described in seven papers authored with former postdocs Alu Cabral and Edu Ferreira. Of special interest are their systematic studies, which revealed the simple parameters (combinations of DSC characteristic temperatures) that best describe glass stability against devitrification (on heating) and its correlation with glass-forming ability (on cooling).

Some other interesting work on the fundamentals of nucleation and growth in undercooled liquids has been carried out with the collaboration of Vlad Fokin and under the competent guidance of theoretician Juern Schmelzer. This team also worked on a new discovery regarding the fact that the crystals in some (supposedly) stoichiometric glasses, such as $1Na_2O.2CaO.3SiO_2$, show very significant departures from stoichiometry. They called these crystals mutant and the related phenomenon the courtyard effect.

More recently, with the help of former postdoc Marcio Nascimento's data-mining skills, the LaMaV team began to examine in detail the diffusion processes that control crystal nucleation, crystal growth, and viscous flow and the controversy regarding the possible breakdown of the Stokes–Einstein (SE) equation at sufficiently low temperatures. The first five papers published on this subject indicated no breakdown of the SE equation from the *liquidus* down to $T \sim 1.2\ T_g$, but at deeper undercoolings below $1.2\ T_g$, there are clear signs of such a breakdown, which has been attributed to spatially heterogeneous dynamics.

A current research topic refers to theoretical estimates and experimental determinations (by NMR, microindentation, and XRD methods) of the *residual stress* fields around crystalline particles in CGs. Here, the significant contributions of Valmor Mastelaro, Oscar Peitl, Joe Zwanziger, and Fran Serbena were fundamental. These stresses sometimes have an enormous impact on the overall mechanical performance of ceramic composites, which include CGs. Five papers have been published on this particular subject.

In 2005, the group started to collaborate with Leon Glebov, Julien Lumeau, and their team at the Creol, University of Central Florida, USA, and have since been working on the very complex crystallization process of the so-called *photo-thermo-refractive* glass (PTRG). Former postdoc Gui Souza and visiting scientist Vlad Fokin have been fundamental in this endeavor. PTRG is a partially crystallized, optical transparent glass that undergoes controlled nanocrystallization after UV exposure followed by double heat treatment. Six papers about PTRG have been published and others are on the way.

Last but not the least, with the key presence of Edu Ferreira, Alu Cabral, Rapha Reis, and Vlad Fokin, LaMaV's team has also been working on the critical assessment of nonisothermal (DSC and DTA) methods to infer crystal nucleation and growth kinetics. Three papers published in the period 2009/2010 clearly indicate that such methods can sometimes be used to probe glass crystallization parameters, but extreme care is required to obtain reliable results. The team concluded that nonisothermal methods often yield misleading results and can be as complex and time consuming as the traditional microscopic and stereological techniques.

Summarizing the above list of original research, Zanotto and collaborators have authored 18 review papers on crystal nucleation, crystal growth, vitrification, and glass crystallization.

Their research work on controlled crystallization of glass has led to the development of several interesting *CGs*, which are described in about 15 papers. These papers discuss new CGs derived from blast furnace and steelmaking slag (hard materials for architecture and construction), sintered CGs that emulate expensive stones such as marble and granite, the first large grain, highly crystalline optically transparent CGs, and a number of bioactive CGs. In this particularly hot subject, the initial help and incitement of the great Larry Hench (the inventor of Bioglass®) and the experimental skills of Oscar Peitl and several dentists, physiotherapists, biologists, and medical doctors have been fundamental. The group has been working on a new highly bioactive CG called BioSilicate® invented, patented, and already licensed to a company (Vitrovita). Their recent papers can be checked using the keyword "Biosilicate" at lamav.weebly.com. About 10 publications on this particular material came out in the period 2009/2010 or are on the way. Finally, a group led by my colleague, Prof. Ana C. M. Rodrigues, has developed some of the most electrically conductive CGs in the world! Several papers have been published on this subject. A recent review paper [Zanotto ED. A bright future for glass-ceramics. *American Ceramic Society Bulletin*. October 2010; 89(12):19–27] recapitulates most of their CG research work.

Conclusions

The above-summarized research comprises a sequential sampling of new insights and significant advances in the understanding of relevant scientific issues related to the nature of glasses and CGs carried out by our team. The group's working philosophy is to focus on a few relevant problems for several consecutive years, at the deepest possible level, ideally until the problems in question have been exhausted or completely solved! Most of the above-listed articles are original and innovative, and lead to significant advances in the body of knowledge about important aspects of phase transformations in glasses. LaMaV's publications demonstrate a strong collaboration with young students and with senior scientists.

In addition to their studies of glass crystallization kinetics and mechanisms, the LaMaV team members have also pursued a number of research projects on rheological, mechanical, biological (Oscar Peitl), electrical (Ana Rodrigues), and other *properties* of glasses and CGs. The team has published about 40 papers on the miscellaneous properties of glass, some of which caused extensive repercussions. For instance, the following paragraph includes an editorial comment published in the *Science* magazine regarding the following two papers, which deal with the alleged viscous flow of window glass at room temperature: (i) Zanotto ED.

Glass Myth Shattered

(*Science Now*, May 16, 1998)

The popular notion that medieval cathedral windows have thickened at the bottom – by slowly flowing like a liquid – doesn't hold water. Even after considering the specific chemical composition of stained glass windows, according to a report in the May issue of the American Journal of Physics, it would take longer than the age of the Universe for the glass to sag appreciably. When materials scientist Edgar Zanotto of the Federal University of São Carlos, Brazil, first heard about sagging medieval windows, "I thought it was just a local [Brazilian] myth". But then he heard the same tale from colleagues in Argentina, and found echoes of it in textbooks and even in the Encyclopedia Britannica. Although glass isn't supposed to flow at room temperature, old glass has many impurities that might help it ooze. So Zanotto sat down to do the calculations. Zanotto looked up the chemical compositions of some 350 medieval glasses and calculated a typical viscosity. The old glass should flow a little more easily than modern glass, he found, but only at temperatures above roughly 200 degrees Celsius (modern glass has to be hotter than about 250 °C to flow). Below 200 °C, the molecules would remain gridlocked. Zanotto also considered an extreme case: germanium oxide glass, which is thought to flow even at bitterly cold temperatures. But even a germanium oxide window would hold its shape, he concluded. Such glass would visibly sag, Zanotto says, "but only if you waited for 10^{32} years." The paper goes a long way to laying the legend of flowing windows to rest, says Jonathan Katz, a materials scientist at the Washington University in St. Louis. Katz, Zanotto, and others think that cathedral glassmakers centuries ago were unable to make even panes and that builders put the fat end on the bottom for stability. If the cathedral glass really is sagging, Katz says, it could be spotted with an interferometer in the lab. "It would be a great graduate student project," he says. Still, there's good evidence the thesis would be a short one: If glass moves at room temperature, he adds, "your camera wouldn't focus right after 10 years on a shelf."

Do cathedral glasses flow? *American Journal of Physics*. 1998; 66(5):392–395 and (ii) Zanotto ED and Gupta PK. Do cathedral glasses flow? Additional Remarks. *American Journal of Physics*. 1999; 67(3):260–262.

Many other scientific journals and newspapers, such as *New Scientist*, *Science News*, *Popular Science*, *The Sciences*, *Scientific American*, *Época*, *Super Interessante*, *Ciência Hoje*, *Reforma*, and another dozen or so, have published editorial comments on these two *AJP* articles.

LaMaV's research has also resulted in some relevant technological developments, as demonstrated by 12 *patents* and numerous services for industry. Most significant is the fact that some leading international glass manufacturers that produce commercial CGs have been using several of the above-listed scientific articles for the development of their products.

Full papers can be downloaded from lamav.weebly.com

Complementary information about glass crystallization and CG microstructures can be found in the following books:

- El-Meliegy E and Van Noort R. *Glasses and Glass Ceramics for Medical Applications*. Springer; 2011. p. 268. ISBN-10 1461412277.

- Höland W and Beall GH. *Glass-Ceramic Technology*. Wiley-Blackwell; 2002. p. 385. ISBN-10 1574981072.

- Strnad Z. *Glass-Ceramic Materials*. 2nd ed. Elsevier Science Ltd.; 1986. p. 268. ISBN-10 0444995242.

- McMillan PW. *Glass-Ceramics*. 2nd ed. Academic Press; 1979. p. 285. ISBN 0-12-485660-8.

- Schulze D. *Sehen, Verstehen, Gestalten*. Werkstoff-Informationsgesellschaft; 1998. In German.

- Vogel W. *Glasfehler*. Springer Verlag; 1993. p. 154. ISBN 3-540-55633-8. In German. http://dx.doi.org/10.1007/978-3-642-58048-2. Accessed 2013 Mar 27.

ACKNOWLEDGMENTS

Si se concibe a la cultura como algo distinto del saber enciclopédico; si se piensa que la cultura es organización, apoderamiento de la propia personalidad, conquista consciente por la cual se comprende el valor y la función en la vida de cada uno, se asume que la cultura es crítica. Desde esta perspectiva, la cooperación se entiende como reconocimiento y apropiación mutua de la cultura del otro.

(UNKNOWN AUTHOR)

I am deeply indebted to all my numerous former and current students, postdocs, and collaborators for the invaluable help and instruction they have afforded us along all these years. The above-described achievements are clearly a result of *collaborative* research work.

Special thanks to Christian Ruessel for hosting several dozen internship students at the famous Otto-Schott-Institut in Jena, Germany, and to Jojo Deubener for hosting some of our postdocs in Clausthal, Germany. To Leon Glebov, Larissa Glebova, Julien Lumeau, Vlad Fokin, Gui Souza, and all the members of the photo-thermo-induced nucleation research team. To Larry Hench, Oscar Peitl, Murilo Crovacce, Marina Trevelin, Juliana Daguano, Clever Chinaglia, Renato Siqueira as well as all the numerous dentists, physiotherapists, biologists, and medical doctors of LaMaV's bio GC team. To Miguel Prado, Ralf Muller, Catia Fredericci, Edu Ferreira, Nora Mora, Anne Barbosa, Vivi Soares, and Rapha Reis—the surface crystallization and sinter crystallization GC team of LaMaV. To Fran Serbena, Mariana Villas-Boas and her internship students of our dental glass-ceramics team, and Leo Gallo of the armor GC group. To Ana Rodrigues, Jean Louis Souquet, and their students—our ionic conducting glass-ceramics team. I am indebeted to Aldo Craievich, Peter James, Mike Weinberg, Don Uhlmann, Vlad Fokin, Juern Schmelzer, Dick Brow, Joe Zwanziger, Valmor Mastelaro, Nora Mora, Paulo Soares, Aluisio Cabral, Marcio Nascimento, Lu Ghussn, Daniel Cassar, Bruno Poletto, Alisson Rodrigues and several other collaborators on fundamental studies on nucleation and crystallization in glasses.

I am also indebted to my former and present colleagues of the *Technical Committee 7* (TC7), the Crystallization and Glass-Ceramic Committee of the International Commission on Glass—who are among the most experienced glass-crystallization scientists in the world—and from whom I have learned a great deal about the intricacies of glass crystallization and the properties and applications of glass-ceramics: Peter James[+], Mike Weinberg[+], Stvan Szabo[+] ([+]deceased), Tadaski Kobubo, Klaus Heide, Wolfgang Pannhorst, Wolfram Höland, Linda Pinckney, Ian Donald, Taka Komatsu, Akihiko Sakamoto, Michael Budd, Maria Pascual, Miguel Prado, Vlad Fokin, Jeri Sestak, Mark Davis, Gilles Querel, Jojo Deubener, Ralf Mueller, Guenter Voelksch, Maria Pascual and Rob Hill. Some outstanding micrographs included in this book have been kindly supplied by Christian Ruessel and the following TC7 members: Vlad Fokin, Miguel Prado, and Taka Komatsu.

Finally, I sincerely thank the Brazilian research-funding agencies Capes, CNPq, and Fapesp for continuously supporting the vitreous materials research group over the past 36 years, the photoshop expert Clever Chinaglia for his valuable help, and Prof. Luiz Nunes Oliveira for proofreading the text.

LETTER FROM S. D. STOOKEY (the Inventor of Glass-ceramics)

October 14th, 2010

Dear Professor Zanotto:

It is a pleasure to congratulate you on the history making paper* you wrote for the new ACERS BULLETIN. I truly believe that, by applying the worldwide knowledge assembled in the article, a new "stone age" can be started, in which, for example, beautiful indestructible buildings can be made from inexpensive materials. Structures can be made that are resistant to fire, corrosion, age, etc.

I hope that the new generation of scientists and engineers will make use of this knowledge and expand it.

Of course I am very grateful for your generosity in giving me credit for the "accidental" discovery.

Best personal wishes, from Donald Stookey (age 95).

*[Zanotto ED. A bright future for glass-ceramics. *American Ceramic Society Bulletin*. October 2010; 89(12):19–27.]

TC 7 Members. PHOTO.
From left to right: Ralf Muller, Guenter Voelksch, Linda Pinckney, Edgar Zanotto, Wolfgang Pannhorst, Takaiuki Komatsu, Miguel Prado, Michael Budd, Joachim Deubener, Wolfram Höland and Ian Donald.
Guests sitting: George Beall and S. D. Stookey.
Courtesy of Mark Davis, Stone Mountain, USA (2006).

I dedicate this book to my crystals Luciana, Bruna, Giulia, and Diva!

CRYSTALS IN GLASS

A Celebration of Science and Art

My fascination with glass crystallization and glass-ceramics (CGs) started about 36 years ago. In January 1977, I had just concluded the Materials Engineering course at the Federal University of São Carlos, Brazil, and was starting my MSc dissertation in Physics at the University of São Paulo, São Carlos, Brazil, under the guidance of Aldo Craievich. My research topic was a small-angle X-ray scattering study of liquid–liquid phase separation and crystal nucleation in Li_2O-SiO_2 glasses. This "model" system is the basis to that used by Stookey in his fortuitous discovery, and for the very successful Fotoceram produced by Corning, as well as of the dental CGs produced by Ivoclar Vivadent—IPS Empress II and e.max. Then I chanced upon Peter McMillan's book on CGs and immediately became enthralled and read it in about a month. It dawned on me that this new type of material could be a great research topic in materials science and engineering, so I followed my insight. A few years later, in 1982, the now late Prof. Peter McMillan—jointly with late Prof. Peter James (my thesis supervisor) and Prof. Harold Rawson—acted as a member of my PhD thesis committee at the Sheffield University, UK. I returned to Brazil to become an adjunct professor at UFSCar and continued with glass crystallization and CG research with the help of a few students. In 1986, out of the blue, I received a phone call from Mike Weinberg—then at Cal Tech, Pasadena, USA—who suggested that we had similar research interests and could collaborate, and invited me to visit his labs in Pasadena. However, this collaboration only *crystallized* by accident in 1987. I was ready to spend a sabbatical year at MIT with Don Uhlmann when, at some stage, he moved to the University of Arizona in Tucson, where I met Mike, who had also recently moved to the same university. We collaborated actively for several years until his untimely death a few years ago. In the period of 1993–2011, we hosted six distinguished guests at LaMaV for a few months to several years: Vlad Fokin, Miguel Prado, Ralf Muller, Juern Schmelzer, Isak Avramov, Jean Louis Souquet, and Fran Serbena. Other very profitable collaborations started 7 years ago with Leon Glebov and Julien Lumeau during my sabbatical at Creol, UCF, Florida, USA, as well as with Joe Zwanziger, Hallifax, Canada; Dick Brow, Rolla, Missouri, USA; and Aldo Boccaccini, Imperial College, London. These hardworking, knowledgeable scientists, together with my two colleagues Oscar Peitl and Ana Rodrigues, and several dedicated former students and postdocs (Edu Ferreira, Catia Fredericci, Nora Diaz, Valmor Mastelaro, Paulo Soares, Alu Cabral, Marcio Nascimento, Gui Souza, Viviane-Soares), and several other new students and postdocs, make up our international *glass crystallization* research network.

My interest on *micrographs* started during my master's research work at the University of São Paulo. This interest was reinforced and grew during my PhD work at the Sheffield University. At Sheffield, I produced about 4000 micrographs (hard prints, since digital photos were not available at that time) of a number of crystals, two of which I have managed to recover and are reproduced in this book.

The driving forces to find, sort out, and compile all these micrographs—which were scattered in numerous different print and electronic files of widely different ages—into a single document are the celebration of *LaMaV's* 36 years; my three decades of participation in the *TC7*, the creation of the *CeRTEV—Center for Research, Technology and Education in Vitreous Materials*—with 16 glass researchers from the two universities in São Carlos (UFSCar and USP); and last but not least, the scientific competence, enthusiasm, and artistic proclivity of our most frequent visitor, *Vladimir Mikailovich Fokin*, who has spent several years at LaMaV and is responsible for producing some spectacular microphotographs in this exhibition.

To commemorate this enjoyable, productive time period, I present a collection of micrographs selected from among thousands that have been taken and analyzed in the past 36 years. I hope this collection of scientific, artistic photographs not only contributes to permanently register and give proper credit to several researchers who participated in important discoveries about glass crystallization but also motivates students and young scientists to join us in this endless but most gratifying *quest to unveil the deeply hidden intricacies of glass crystallization and the resulting properties!*

6

INTERNAL NUCLEATION IN GLASSES

Crystals in Glass: A Hidden Beauty, First Edition. Edgar D. Zanotto.
©2013 John Wiley & Sons, Inc. and The American Ceramic Society. Published 2013 by John Wiley & Sons, Inc.

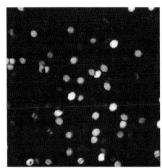

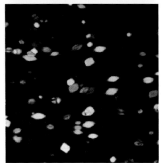

Lithium Disilicate Crystals in an Isochemical Glass

$Li_2O.2SiO_2$ glass is one of the few that shows internal nucleation without the aid of nucleating agents and is considered the "*Drosophila*" of glass crystallization studies. After a single heat treatment above T_g, a size distribution of crystal sizes results due to simultaneous nucleation and growth. The two similar micrographs on the left-hand side—taken by different researchers using different glass melts at widely different times—show the great reproducibility afforded by such type of studies. The largest crystals are about 50 μm. This crystal phase is the basis of several commercial glass-ceramics.

Optical microscopy, polarized transmitted light.

 Sheffield University (UK) and LaMaV (UFSCar)

 Upper: Ed Zanotto (1980); Lower: Vlad Fokin (1998)

Zanotto ED and James PF. Experimental tests of the classical nucleation theory for glasses. *Journal of Non-Crystalline Solids*. 1985; 74(2–3):373–394. http://dx.doi.org/10.1016/0022-3093(85)90080-8.

Fokin VM, Zanotto ED, Yuritsyn NS and Schmelzer JWP. Homogeneous crystal nucleation in silicate glasses: A 40 years perspective. *Journal of Non-Crystalline Solids*. 2006; 352(26–27):2681–2714. http://dx.doi.org/10.1016/j.jnoncrysol.2006.02.074. DOI: 10.1016/j.jnoncrysol.2006.02.074.

9

50 μm

Spherulitic Crystals in a Stoichiometric Barium Disilicate Glass

These internal 20 μm $BaO.2SiO_2$ crystals growing in an isochemical glass are *spherulitic*; that is, they have a residual glass phase embedded within the fine crystal branches of the spherical crystal-like regions. XRD experiments demonstrated for the first time that each region is only about 65% crystalline; that is, they are spherulitic. Thus, spherulites can indeed nucleate and grow in inorganic glasses; not only in polymers, as previously thought.

Optical microscopy, transmitted light.

 Sheffield University (UK) and LaMaV (UFSCar)

 Ed Zanotto (1980)

Zanotto ED and James PF. Experimental test of the general theory of transformation kinetics: Homogeneous nucleation in a $BaO.2SiO_2$ glass. *Journal of Non-Crystalline Solids*. 1988; 104(1):70–72. http://dx.doi.org/10.1016/0022-3093(88)90183-4.

11

20 µm

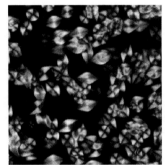

Internal Crystallization in Ti-cordierite Glass

Pure stoichiometric cordierite ($2MgO.2Al_2O_3.5SiO_2$) glass undergoes surface nucleation, but the same glass doped with >6 mol% TiO_2 shows internal crystallization of μ-cordierite. Several crystal morphologies can appear depending on the time and temperature of heat treatment.

Cordierite is a major crystal phase in several types of glass-ceramics.

Optical microscopy, polarized light.

 LaMaV (UFSCar) and TC7

 Vlad Fokin (1998)

Fokin VM and Zanotto ED. Surface and volume nucleation and growth in TiO_2-cordierite glasses. *Journal of Non-Crystalline Solids*. 1999; 246(1):115–127. http://dx.doi.org/10.1016/S0022-3093(99)00007-1.

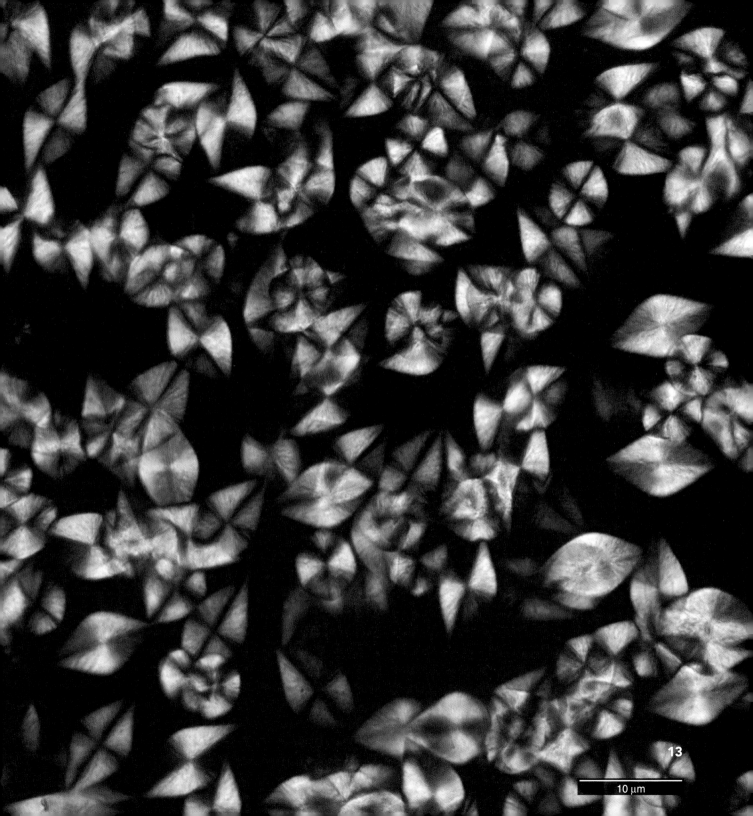

13

10 μm

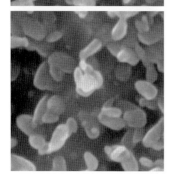

Papaya-seed-like Nanocrystals in Fresnoite Glass

These fresnoite ($2BaO.TiO_2.2SiO_2$) crystals copiously nucleate and grow in the interior of an isochemical fresnoite glass. They display the highest nucleation rates ever reported for an oxide glass, $\sim 10^{17}/m^3.s$. The three micrographs demonstrate the effect of thermal treatment; the longer the more crystals appear, the larger they become.

Scanning electron microscopy.

 LaMaV (UFSCar)

 Alu Cabral and Clever Chinaglia (2002)

Cabral AA, Fokin VM, Zanotto ED and Chinaglia CR. Nanocrystallization of fresnoite glass. I. Nucleation and growth kinetics. *Journal of Non-Crystalline Solids*. 2003; 330(1–3):174–186. http://dx.doi.org/10.1016/j.jnoncrysol.2003.08.046.

Cabral AA, Fokin VM and Zanotto ED. Nanocrystallization of fresnoite glass. II. Analysis of homogeneous nucleation kinetics. *Journal of Non-Crystalline Solids*. 2004; 343(1–3):85–90. http://dx.doi.org/10.1016/j.jnoncrysol.2003.08.046.

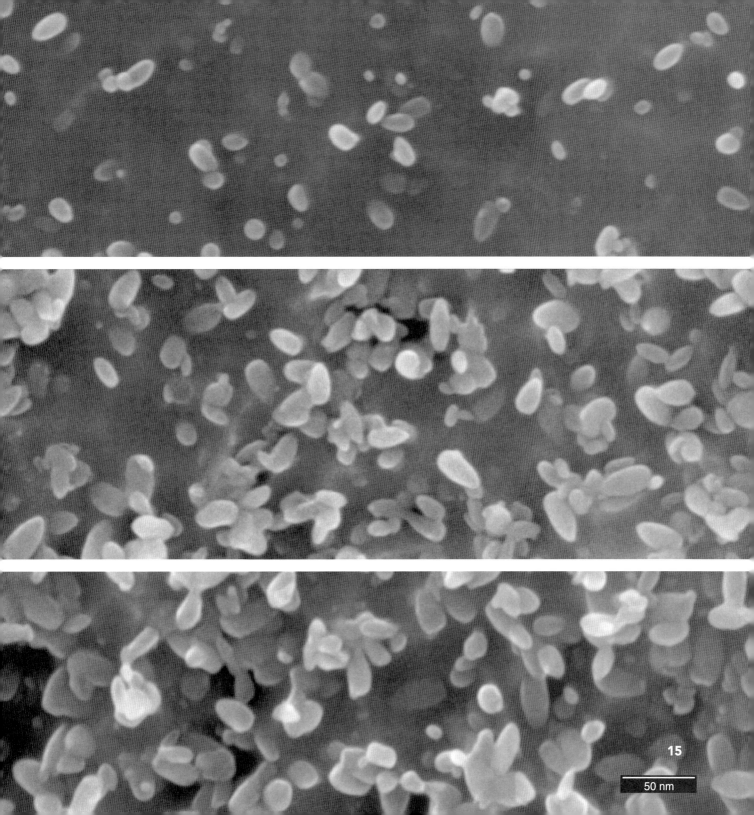

15

50 nm

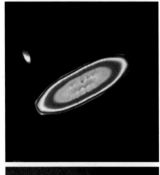

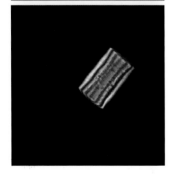

Lithium Diborate Crystals in an Isochemical Glass

Lithium diborate glass shows internal nucleation without the aid of nucleating agents, and several crystal morphologies can appear depending on the heat treatment. After a single-stage heat treatment above T_g, very rare Saturn-ring-like and cylindrical $Li_2O.2B_2O_3$ crystals of about 50 µm develop.

Optical microscopy, polarized transmitted light.

LaMaV (UFSCar)

Lu Ghussn (2008) and Dani Cassar (2011)

Ghussn L. Post Doctoral Report Submitted to FAPESP (2008).
Supervisor: Zanotto, E. D.

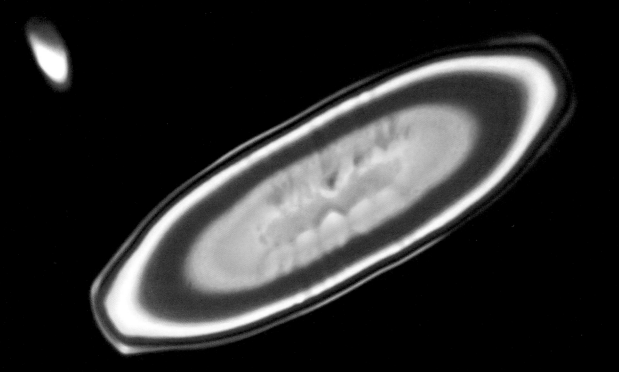

17

5 μm

Internal Crystal in a Diopside Glass

Surface nucleation is the hallmark of diopside ($CaO.MgO.2SiO_2$) glass. This micrograph shows a very rare case (the only one ever reported) of a diopside crystal in the interior of a diopside glass. Volume crystallization was possibly caused by heterogeneous nucleation on some solid impurity particle or by a fortuitous homogeneous nucleation event.

Optical microscopy polarized transmitted light.

 LaMaV (UFSCar)

 Vlad Fokin (2009)

19

50 mm

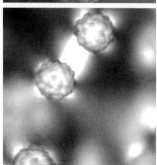

Lithium Niobium Disilicate (Double) Crystals in a Nonstoichiometric Glass

After a double-stage heat treatment of $45Li_2O\text{-}20Nb_2O_5\text{-}35SiO_2$ glass above T_g, a narrow size distribution of mixed $Li_2O.SiO_2\text{-}Li_2O.Nb_2O_5$ crystals of about 300 μm grow in the glass interior. Cracks are generated due to the thermal expansion mismatch between the glassy matrix and the crystals. The lower photo in the left-hand side shows the same crystals with a *sweetsop-like* shape.

Optical microscopy, reflected light.

 LaMaV (UFSCar)

 Vivi Oliveira and Mu Crovacce (2009)

Souza LA. Current problems on the crystallization of borate, silicate and telluride glasses. [Tese]. Universidade Estadual Paulista; 2004.

Supervisors: Zanotto, E. D. and Rodrigues, A.C.M.

21

60 μm

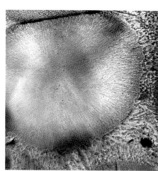
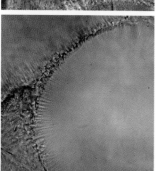

Crystals in Li$_2$O-Doped Soda-lime-silica Glasses

Fully crystallized 1soda-2lime-3silica glasses with 5.6 mol% Li$_2$O and 7.5 mol% Li$_2$O. The large particle is the 123 phase. A second (unknown) phase, which is rich in Li, Ca, and Si, precipitates in the grain boundary.

Optical microscopy, reflected light.

 LaMaV (UFSCar)

 Tiago Mosca (2009)

Poster presented at the II Summer School of Materials Physics, IFSC-USP, São Carlos (2009).

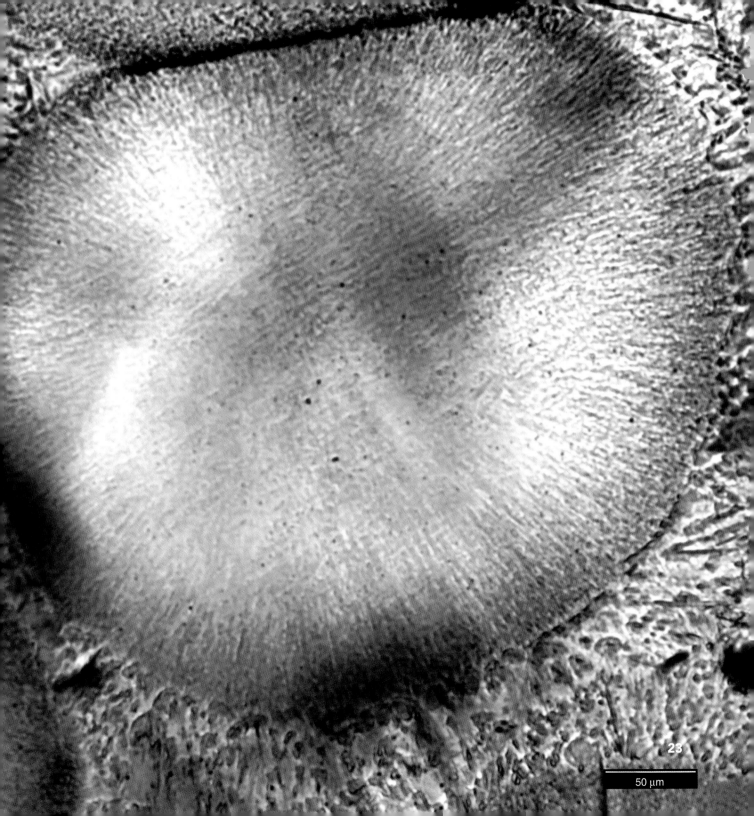

23

50 μm

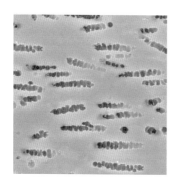

Textured Worm-like Crystals in a Bioactive Glass Fiber

$KCaPO_x$ crystals in the interior of a K-Na-Ca-Mg-Si-P-B-O bioactive glass fiber. The distinctive feature here is the elongated crystals aligned in the fiber-drawing direction, which can lead to anisotropic properties such as those of natural bone.

Scanning electron microscopy.

 LaMaV (UFSCar)

 Mu Crovacce and Ma Trevelin (2011)

Souza MT. Desenvolvimento de fibras de vidro altamente bioativos. [Dissertação] (Development of highly bioactive glass fibers. [Master´s dissertation]. São Carlos: Universidade Federal de São Carlos; 2011.

Supervisors: Zanotto, E. D. and Peitl, O.

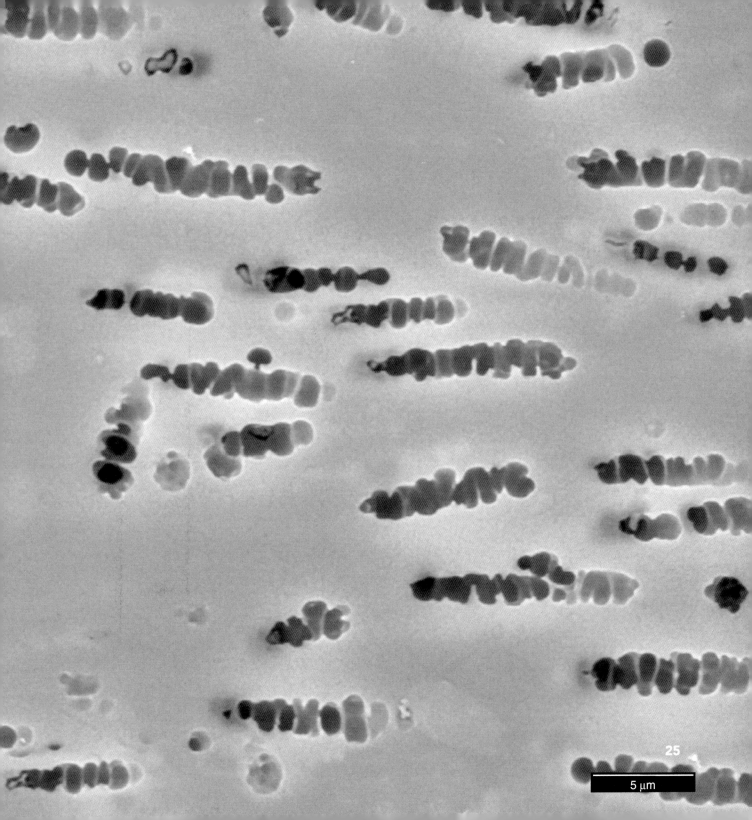

25

5 µm

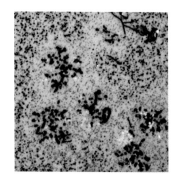

Liquid–liquid Phase Separation and Crystallization in Photo-thermo-refractive Glass

Photo-thermo-refractive (PTR) glass contains below 1% by volume of nanosized NaF crystals uniformly aligned forming diffraction gratings. Crystallization is induced by exposure to UV light followed by heat treatment just above the glass-transition temperature, T_g. In this micrograph, darker agglomerates are dendritic NaF crystals embedded in a vitreous matrix filled with amorphous droplets resulting from liquid phase separation, which are rich in silica. In this particular case, to reveal the later stages of crystallization, the microstructure was hyperdeveloped by heat treatment at higher temperatures. This material is marketed by the US company OptiGrate for numerous optical applications.

Optical microscopy, reflected light.

 Creol-UCF (USA) and LaMaV (UFSCar)

 Gui Souza and Vlad Fokin (2010)

Souza GP, Fokin VM, Rodrigues CF, Rodrigues ACM, Zanotto ED, Lumeau J et al. Liquid-liquid phase separation in photo-thermo-refractive glass. *Journal of the American Ceramic Society*. 2011; 94:145–150. http://dx.doi.org/10.1111/j.1551-2916.2010.04053.x.

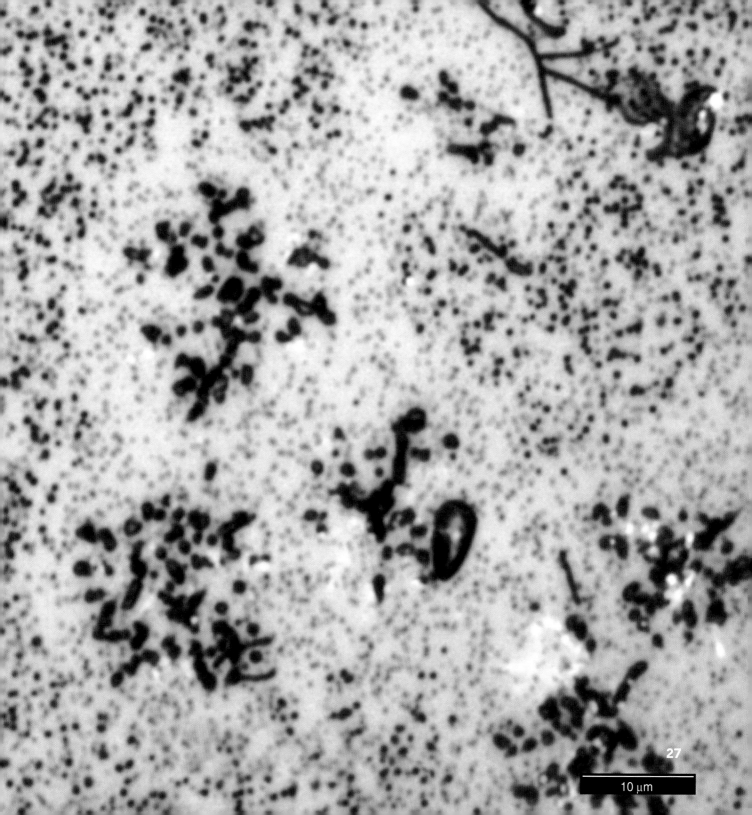

27

10 µm

Star-like Crystals in the Volume of PTR Glass

NaF crystals in the interior of a PTR glass. Treatment at a high temperature near the solubility limit. UV-unexposed, heat-treated at 700 °C for 30 min.

Optical microscopy, transmitted light.

Awarded best micrograph of the Second Summer School on Materials Physics, IFSC, USP (2009).

 Creol-UCF (USA), LaMaV (UFSCar) and TC7

 Carlos Baptista, Vlad Fokin, and Gui Souza (2009)

Souza GP, Fokin VM, Zanotto ED, Lumeau J, Glebova L and Glebov LB. Micro and nanostructures in partially crystallised photothermorefractive glass. *Physics and Chemistry of Glasses: European Journal of Glass Science and Technology, Part B.* 2009; 50(5): 311–320.

Lumeau J, Sinitskii A, Glebova L, Glebov LB and Zanotto ED. Spontaneous and photo-induced crystallisation of photo-thermo-refractive glass. *Physics and Chemistry of Glasses: European Journal of Glass Science and Technology, Part B.* 2007; 48(4):281–284.

29

10 μm

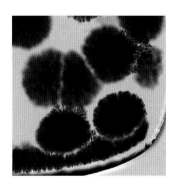

Cristobalite Crystals in PTR Glass

At more advanced stages, after precipitation of the desired NaF crystals, spherulitic cristobalite crystals appear in PTR glass. In this case, the glass was not exposed to UV and was hyperdeveloped by extensive thermal treatment.

Optical microscopy, transmitted light.

 Creol-UCF (USA), LaMaV (UFSCar) and TC7

 Vlad Fokin and Gui Souza (2009)

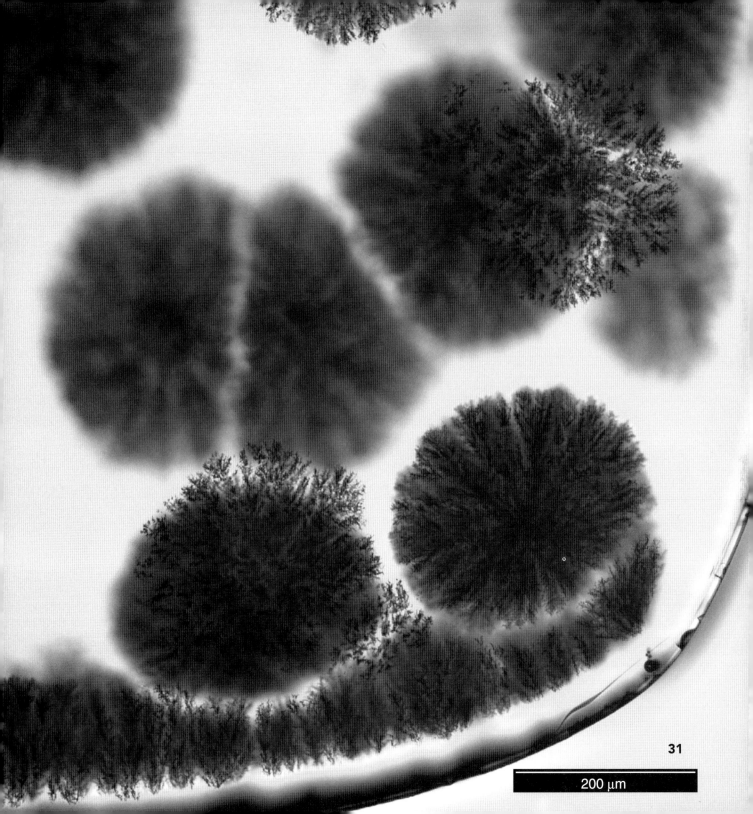

31

200 μm

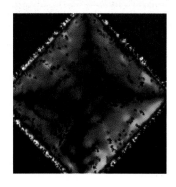

Surface Layer and Internal Crystallization in PTR Glass

NaF crystals on the surface layer and in the glass volume. PTR glass with 50% of the typical bromine content, UV-unexposed, hyperdeveloped at 600 °C for 24 h.

Optical microscopy, transmitted, cross-polarized light.

 Creol-UCF (USA), LaMaV (UFSCar), and TC7

 Gui Souza and Vlad Fokin (2009)

CRYSTALS IN GLASS: A Hidden Beauty

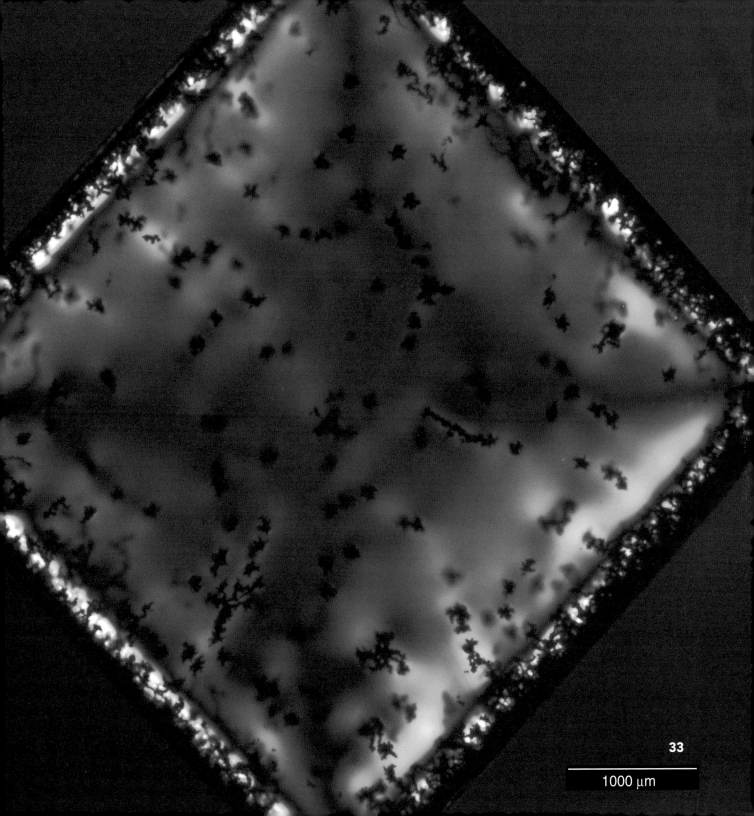

1000 μm

33

The Courtyard Effect in Stoichiometric Soda-lime-silica Glass

Sodium-rich $Na_{2+x}O.2Ca_{1-x}O.3SiO_2$ crystals in the interior of a 1-2-3 stoichiometric glass. The larger crystals were produced by a first heat treatment and the smaller by a second one. The smaller crystals could not nucleate and grow in the Na-depleted "courtyards" of the larger crystals.

Optical microscopy, reflected light.

  LaMaV (UFSCar)

Vlad Fokin (1998)

Fokin VM, Potapov OV, Zanotto ED, Spiandorello FM, Ugolkov VL and Pevzner BZ. Mutant crystals in $Na_2O.2CaO.3SiO_2$ glasses. *Journal of Non-Crystalline Solids*. 2003; 331(1–3): 240–253. http://dx.doi.org/10.1016/j.jnoncrysol.2003.08.074.

Fokin VM, Potapov OV, Chinaglia CR and Zanotto ED. The effect of pre-existing crystals on the crystallization kinetics of a soda-lime-silica glass: The courtyard phenomenon. *Journal of Non-Crystalline Solids*. 1999; 258(1):180–186. http://dx.doi.org/10.1016/S0022-3093(99)00417-2.

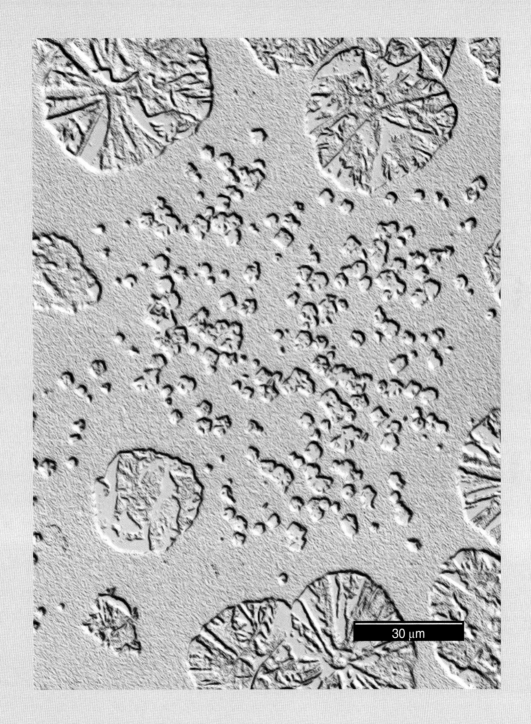

30 μm

The Courtyard Effect in Stoichiometric Soda-lime-silica Glass

Sodium-rich $Na_{2+x}O.2Ca_{1-x}O.3SiO_2$ crystals in the volume of a 1soda-2lime-3silica stoichiometric glass. The larger crystals were produced by a first heat treatment (T_1) and the smaller in a second treatment ($T_2 < T_1$). The smaller crystals could not nucleate and grow in the Na-depleted "courtyards" of the large crystals.

The inset shows a sample treated only at T_2.

Optical microscopy, reflected light.

 LaMaV (UFSCar)

 Vlad Fokin (2008)

Rodrigues ACM, Niitsu GT, Zanotto ED, Prado MO and Fokin VM. Erratum to "Crystallization kinetics of 1Na$_2$O.2CaO.3SiO$_2$ glass monitored by electrical conductivity measurements" [J. Non-Cryst. Solids 353 (2007) 2237–2243]. *Journal of Non-Crystalline Solids*. 2008; 354(26):3098–3107. http://dx.doi.org/10.1016/j.jnoncrysol.2008.01.001.

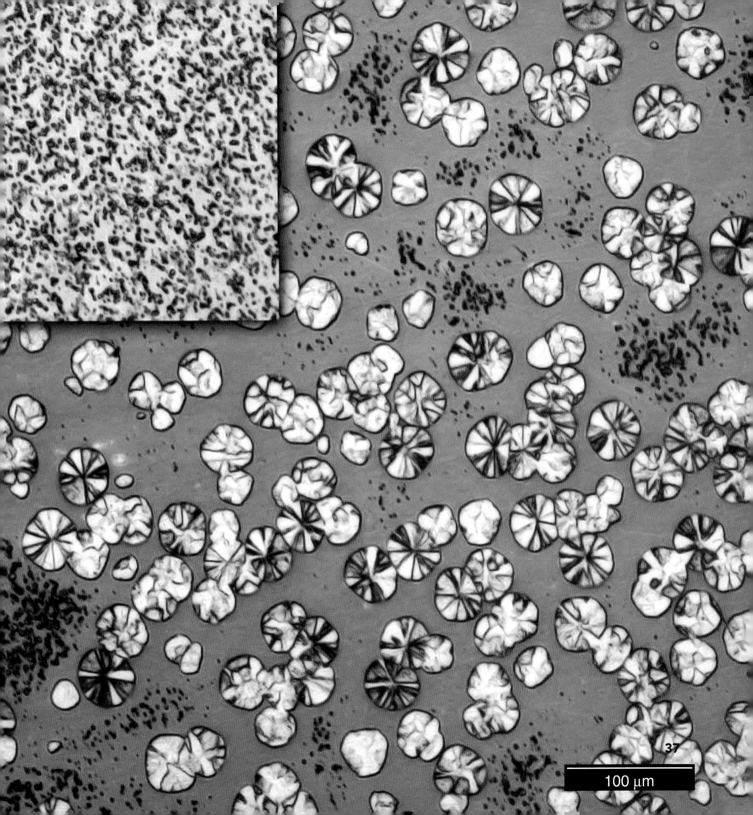

100 μm

The Courtyard Effect—LS Crystals in a Eutectic Glass

Lithium metasilicate (LS) crystals in a cross section parallel to the surface of a $CaO-Li_2O-SiO_2$ eutectic glass. The largest crystal was nucleated in the glass interior, whereas the small crystals nucleated and grew from the sample surface toward the center. The small crystals do not touch each other and could not nucleate and grow in the Li-depleted "courtyards" of the largest crystal.

Optical microscopy, reflected light.

 LaMaV (UFSCar) and TC7

 Vlad Fokin (2011)

39

10 µm

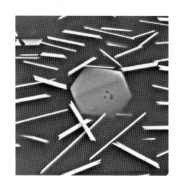

Hematite Crystals in Soda-lime-silica Glass

Soda-lime-silica glasses heavily doped with 25 mol% Fe_2O_3 show internal crystallization of hematite. The micrograph shows hematite crystals in different orientations.

These materials are being tested for hyperthermia treatment of tumors.

Scanning electron microscopy.

 Otto-Schott-Institut (Jena, Germany)

 Wolfgang Wisniewski (2010). Courtesy of Christian Ruessel

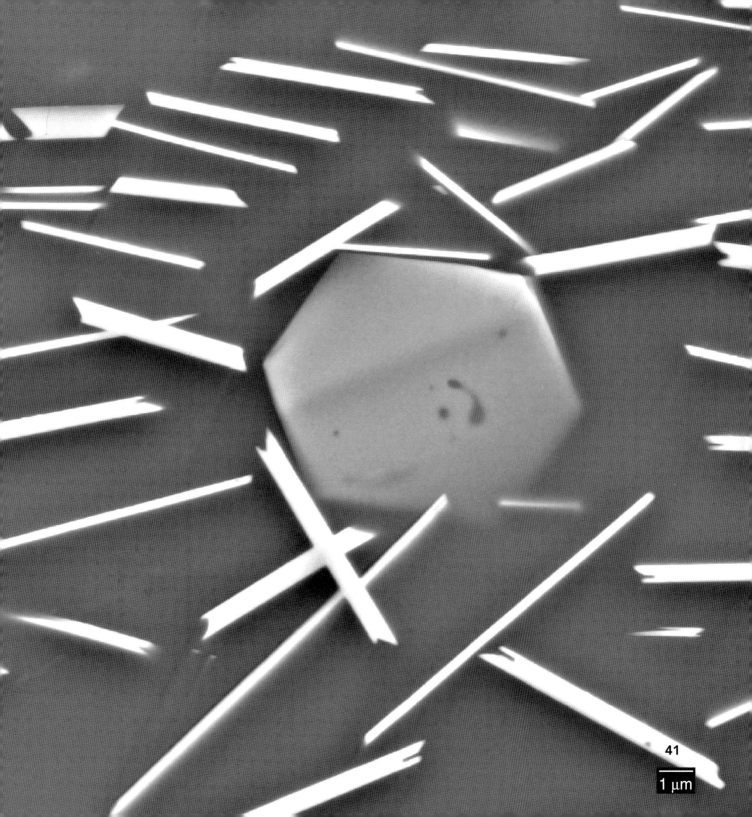

41

1 μm

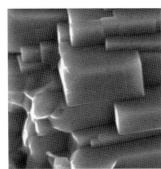

Ionic Conducting Glass-ceramics

Totally crystallized Li^{+1} conducting glass-ceramics. With heat treatment, NASICON-type ($Li_{1,5}Al_{0,5}Ge_{1,5}(PO_4)_3$) crystals appear fully embedded, with unusual grain boundaries and geometry in the mosaic form.

This material is being developed for use in solid-state batteries.

Scanning electron microscopy.

 LaMaV (UFSCar)

 Zé Semanate and Clever Chinaglia (2011)

43

0.5 μm

SURFACE NUCLEATION ON GLASSES

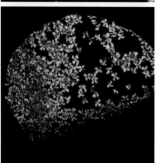

Surface Crystallization of Lithium Diborate Glass

Lithium diborate glass shows both internal and surface nucleation. Several crystal morphologies can appear in this glass depending on the heat treatment. Very rare butterfly-like $Li_2O.2B_2O_3$ crystals of about 50 μm develop after heating above T_g.

Optical microscopy, polarized transmitted light.

 LaMaV (UFSCar)

 Lu Ghussn (2008)

Ghussn L. Post Doctoral Report Submitted to FAPESP (2008).

Supervisor: Zanotto, E. D.

47

25 µm

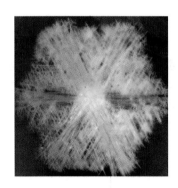

Cordierite Crystal on the Surface of a Cordierite Glass

Most crystals in glasses have a well-defined external shape (hexagons, in this case) and look like single crystals under an optical microscope. But this micrograph shows that, under the power of an electron microscope, some crystals show a very fine dendritic internal structure with a residual glassy phase between the crystal arms.

Scanning electron microscopy.

 Otto-Schott-Institut (Jena, Germany)

 Kittel T (2008). Courtesy of Christian Ruessel

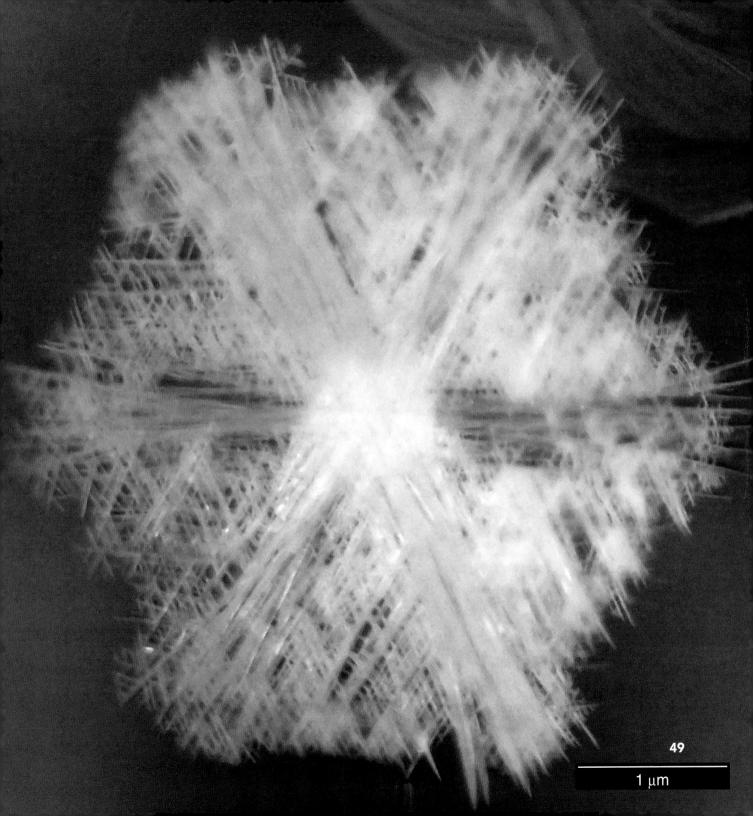

49

1 µm

Surface Nucleation on Cordierite Glass

Cordierite crystal nucleating on a solid (impurity) particle on the glass surface. The largest axis has about 30 μm. This study was coordinated by the TC7 (*Crystallization and Glass-Ceramics Committee of the International Commission on Glass*) to unveil the secrets of surface crystallization.

Scanning electron microscopy.

 LaMaV (UFSCar) and TC7

 Ervino Ziemath and Nora Mora (1992)

Ziemath EC, Diaz-Mora N and Zanotto ED. Crystal morphologies on a cordierite glass surface. *Physics and Chemistry of Glasses.* 1997; 38(1):1–5.

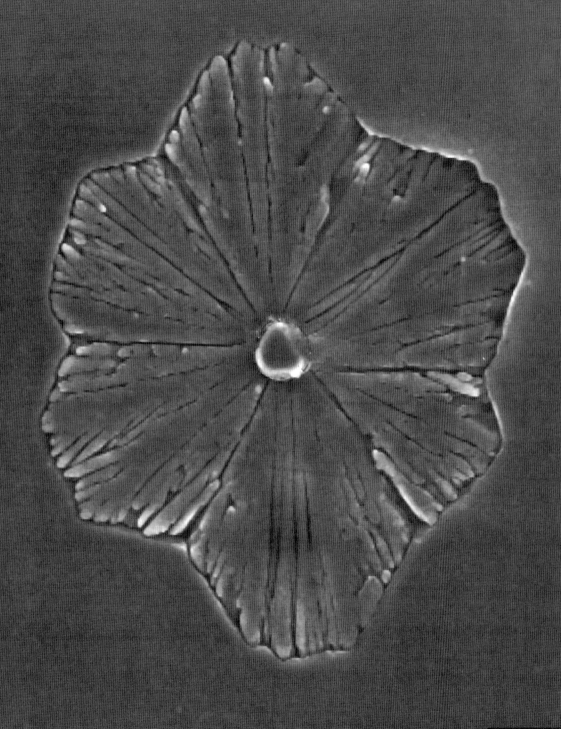

51

10 μm

Nucleation on Scratches, Cracks, and Bubbles

Upper left: preferential nucleation on a scratch in a $BaO.2SiO_2$ glass. Upper right: preferential nucleation on a bubble surface in a $BaO.2SiO_2$ glass. Lower: preferential nucleation of diopside crystals ($CaO.MgO.2SiO_2$) along a scratch on the surface of a diopside glass.

Optical microscopy, reflected light.

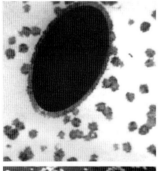

 LaMaV (UFSCar) and TC7

 Ed Zanotto (1981)

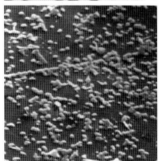

Zanotto ED. Surface nucleation in a diopside glass. *Journal of Non-Crystalline Solids*. 1991; 130:217–219. http://dx.doi.org/10.1016/0022-3093(91)90458-I.

Müller R, Zanotto ED and Fokin VM. Surface crystallization of silicate glasses: Nucleation sites and kinetic. *Journal of Non-Crystalline Solids*. 2000; 274(1):208–231. http://dx.doi.org/10.1016/S0022-3093(00)00214-3.

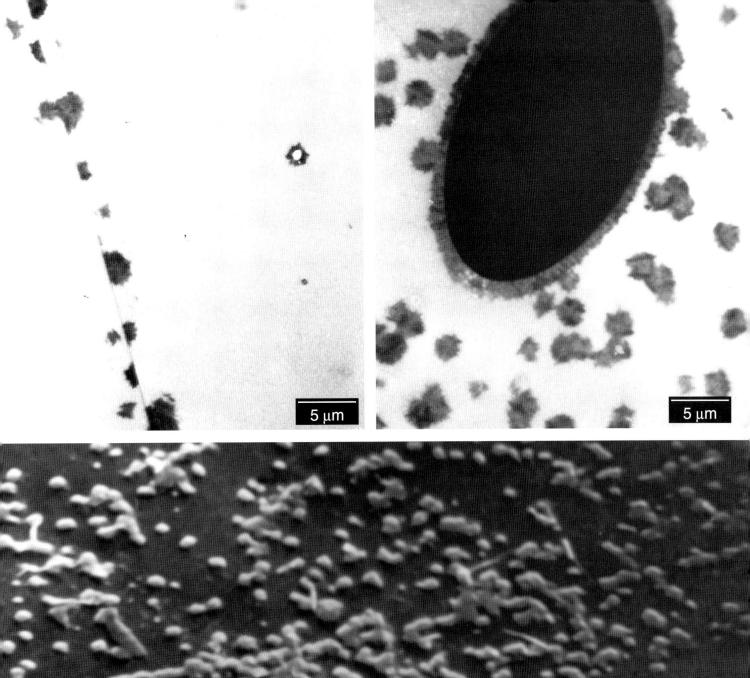

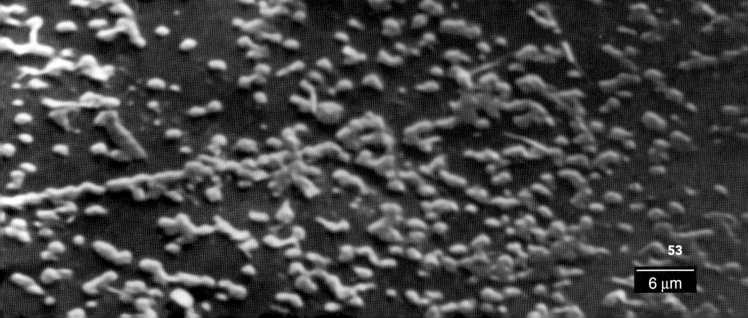

53

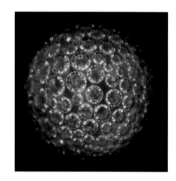

Crystals on Bubble Surfaces in a Diopside Glass

Diopside ($CaO.MgO.2SiO_2$) crystals—white spots—on the surface of bubbles in a diopside glass sphere.

Optical microscopy, transmitted light.

 CNEA (Argentina) and LaMaV (UFSCar)

 Rapha Reis and Dani Cassar (2010)

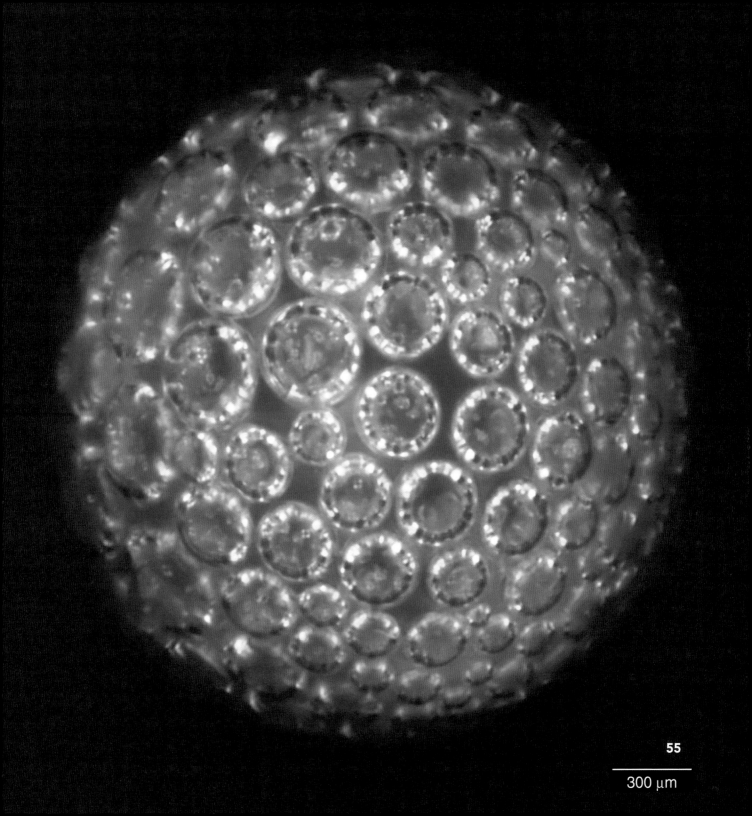

55

300 μm

Surface Crystallization on a Calcium Phosphate Glass

The very narrow crystal size distribution indicates fast heterogeneous nucleation on preexisting sites on the surface of a calcium phosphate glass. All the calcium phosphate crystals were nucleated and started to grow at the same time.

Optical microscopy, reflected light.

 LaMaV (UFSCar)

 Lu Ghussn (2008)

57

10 µm

Surface Crystallization on Ca-rich Diopside Glass

Devitrification on the cooling path at high temperatures leads to large (millimeter), well-defined crystals. The lath-shaped yellow-brown crystals are optically active and thus refer to a different phase than the pyramidal gray crystals.

Optical microscopy, polarized light.

 LaMaV (UFSCar)

 Du Ferreira and Rapha Reis (2007)

59

1 mm

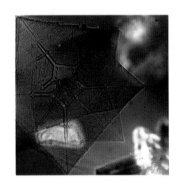

Surface Crystallization on Ca-rich Diopside Glass

Devitrification on the cooling path at high temperatures leads to large (millimeter), well-defined crystals. The lath-shaped yellow-brown crystals are optically active and thus refer to a different phase than the pyramidal gray crystals.

Optical microscopy, polarized light.

 LaMaV (UFSCar)

 Du Ferreira and Rapha Reis (2007)

61

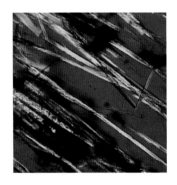

Wollastonite Needles in a Commercial Window (Soda-lime-silica) Glass

These wollastonite ($CaO.SiO_2$) crystals resulted from spontaneous crystallization (devitrification) during glass fabrication. This is a common defect often found in bottle and window glasses.

Optical microscopy, polarized transmitted light.

 LaMaV (UFSCar)

 Du Ferreira (2003)

50 μm

Needle-like Crystals on CaO-Li$_2$O-SiO$_2$ Glass

Volume crystallization between the solidus and the liquidus. Needle-like wollastonite crystals appear on the cooling path of a eutectic glass.

Optical microscopy, transmitted light.

 LaMaV (UFSCar)

 Vlad Fokin (2011)

65

50 μm

"Onion-rings" $1Na_2O.2CaO.3SiO_2$ Crystals on the Surface an Isochemical Glass

These crystals were produced by a series of intermittent heating–cooling cycles, which leave clear marks on them and facilitate crystal-growth rate measurements. Another special finding here is the faster growth rates of the surface crystals when compared with the same crystals in the glass interior (ghost images below the surface).

Optical microscopy, reflected light.

 Institute of Silicate Chemistry (Russia)

 Courtesy of Nico Yuritsyn (2005)

Yuritsyn N. Crystal growth on the surface and in the bulk of $Na_2O.2CaO.3SiO_2$ glass. In: Schmelzer J, editor. *Nucleation Theory and Applications*. Wiley-VCH; 2005. p. 22–42.

67

10 µm

Laser-induced Surface Crystallization of Sm_2O_3-Bi_2O_3-B_2O_3 Glass

Certain glasses can be selectively crystallized under a powerful laser beam. This micrograph shows a bird-like $Sm_xBi_{1-x}BO_3$ single crystal that emulates the famous hieroglyphs of Nazca, Peru. The longest axis is about 1.2 mm.

Optical microscopy, transmitted light.

 Nagaoka University (Japan) and TC7

 Taka Komatsu et al., reproduced by permission of Elsevier

240 µm

VISCOUS SINTERING WITH CONCURRENT CRYSTALLIZATION

Sintering with Concurrent Surface Crystallization of Diopside Glass Spheres

Neck formation during viscous sintering of diopside glass particles with diopside crystals on the glass particle surfaces, which hinder sintering.

Scanning electron microscopy.

 LaMaV (UFSCar)

 Anne Barbosa and Cel Goulart (2011)

Reis RMCV. Glass sintering with concurrent crystallization. [MSc dissertation]. São Carlos: Universidade Federal de São Carlos; 2009.

Supervisor: Zanotto, E. D.

73

25 µm

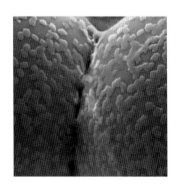

Sintering with Concurrent Crystallization of Two Diopside Glass Spheres

Formation of a neck in the early stages of viscous sintering of two glass particles of diopside, with the presence of many crystals on their surface that retard sintering.

Scanning electron microscopy.

 LaMaV (UFSCar)

 Vivi Soares and Rapha Reis (2009)

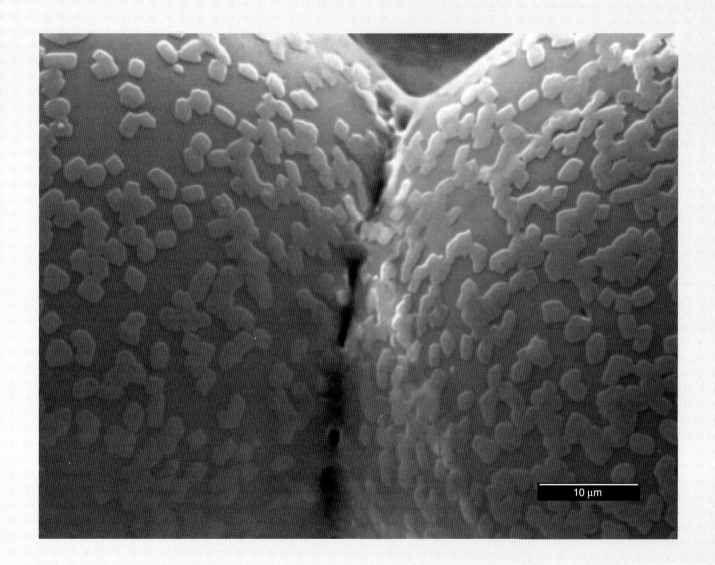

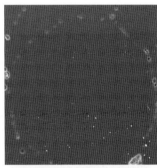

Sintering and Surface Crystallization of Spherical Soda-lime-silica Glass Particles

The upper micrograph reveals "defect" generation at the sintering fronts on the original glass particle surfaces. The second micrograph—a zoom of one "defect"—shows that they are crystals that nucleated on the particles' interfaces.

Optical microscopy, reflected light.

 LaMaV (UFSCar)

 Cat Fredericci and Mig Prado (2002)

Prado MO, Fredericci C and Zanotto ED. Isothermal sintering with concurrent crystallization of polydispersed soda-lime-silica glass beads. *Journal of Non-Crystalline Solids*. 2003; 331(1–3):145–156. http://dx.doi.org/10.1016/j.jnoncrysol.2003.08.076.

Prado MO, Fredericci C and Zanotto ED. Non-isothermal sintering with concurrent crystallization of polydispersed soda-lime-silica glass beads. *Journal of Non-Crystalline Solids*. 2003; 331(1–3):157–167. http://dx.doi.org/10.1016/j.jnoncrysol.2003.08.077.

50 μm

77

2,5 μm

EUTECTIC CRYSTALLIZATION

Crystals in Glass: A Hidden Beauty, First Edition. Edgar D. Zanotto.
©2013 John Wiley & Sons, Inc. and The American Ceramic Society. Published 2013 by John Wiley & Sons, Inc.

Crystallization Propagating from the Surface of a CaO-Li$_2$O-SiO$_2$ Glass

Lithium metasilicate crystals that nucleated on the surface and are growing toward the sample center. Orthogonal cross section. This micrograph resembles Vincent Van Gogh's famous painting "Wheat Field and a Bird (1887)."

Optical microscopy, reflected light.

 LaMaV (UFSCar) and TC7

 Vlad Fokin (2011)

81

100 μm

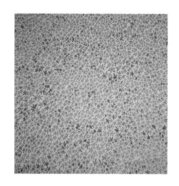

Eutectic Crystallization on a CaO-Li$_2$O-SiO$_2$ Glass

This micrograph of a cross section of the surface layer shows a very interesting phenomenon; there are thousands of lithium metasilicate crystals but none touch each other! This is due to the existence of Li-depleted diffusion layers around each crystal in another example of the courtyard effect.

Optical microscopy, reflected light.

 LaMaV (UFSCar) and TC7

 Vlad Fokin (2011)

83

100 μm

Eutectic Crystallization of CaO-Li$_2$O-SiO$_2$ Glass

These intricate morphology eutectic crystals—common in metallic alloys—appeared in a glass sample after heat treatment of a CaO-Li$_2$O-SiO$_2$ glass somewhat below the *solidus*.

Optical microscopy, reflected light.

 LaMaV (UFSCar) and TC7

 Vlad Fokin (2011)

85

25 μm

Hummingbird-like Crystals on the Surface of a Eutectic CaO-Li$_2$O-SiO$_2$ Glass

Volume nucleation of lithium metasilicate crystals of about 200 μm grown by double-stage heating of a CaO-Li$_2$O-SiO$_2$ glass. They resemble hummingbirds.

Optical microscopy, polarized transmitted light.

 LaMaV (UFSCar) and TC7

 Vlad Fokin (2011)

87

100 μm

Orchid-like Crystallization in a Eutectic CaO-Li$_2$O-SiO$_2$ Glass

Volume nucleation of orchid-like lithium metasilicate crystals grown by double-stage heating. They show internal cracks.

Optical microscopy, polarized transmitted light.

 LaMaV (UFSCar) and TC7

 Vlad Fokin (2011)

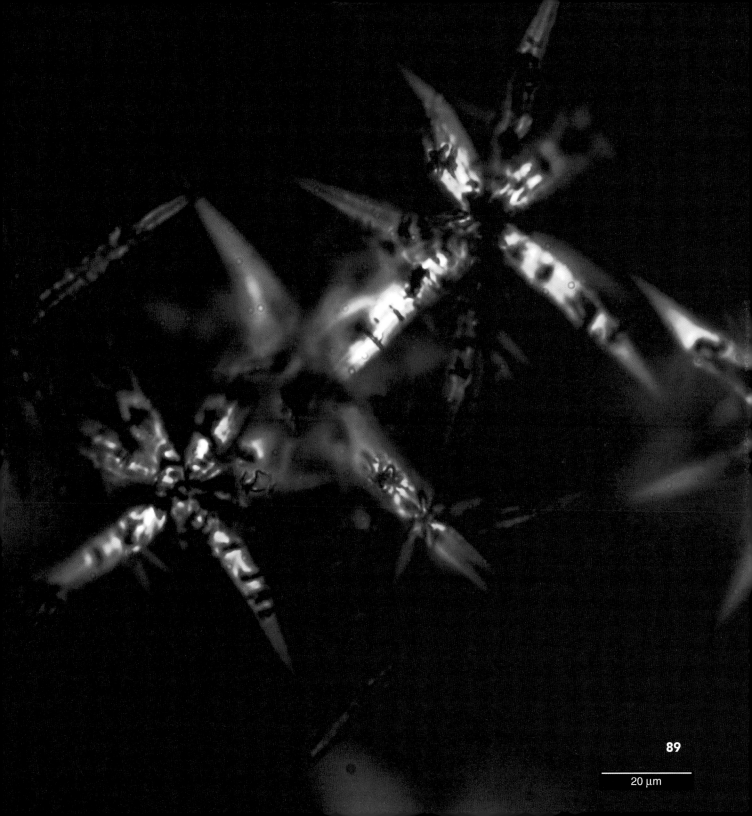

89

20 µm

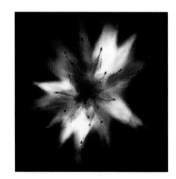

Star-fruit-like Crystals in a Eutectic Glass

Surface nucleation of lithium metasilicate growing toward the interior of a $CaO\text{-}Li_2O\text{-}SiO_2$ glass. They resemble star fruits.

Optical microscopy, polarized transmitted light.

 LaMaV (UFSCar) and TC7

 Vlad Fokin and Dani Cassar (2011)

91

20 μm

CRACKS AND BUBBLES IN GLASS-CERAMICS

Self-cracking of Crystals in Isochemical Glass

Due to their elastic and thermal mismatch, crystals in glasses always produce a residual stress field. Cracks may spontaneously generate when the crystals reach a critical size.

This is the case shown for lithium disilicate crystals growing in an isochemical glass.

Optical microscopy, transmitted light.

 LaMaV (UFSCar)

 Val Mastelaro and Ed Zanotto (1998)

Mastelaro VR and Zanotto ED. Anisotropic residual stresses in partially crystallized Li_2O-$2SiO_2$ glass-ceramics. *Journal of Non-Crystalline Solids*. 1999; 247:79–86. http://dx.doi.org/10.1016/S0022-3093(99)00038-1.

95

40 µm

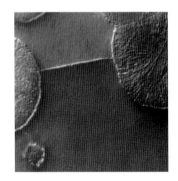

Spontaneous Crack Propagation in a Bioactive Glass-ceramic

Crystals embedded in glasses always produce a residual stress field. When they reach a certain critical size, cracks may be spontaneously generated.

In this particular case of a highly bioactive glass-ceramic, cracks propagate within the residual glass phase and are deflected by the crystals, which enhance the material's toughness and strength. Currently, this material is being clinically tested in humans for replacement of small bones such as middle-ear bones.

Optical microscopy, reflected light.

 University of Florida (USA) and LaMaV (UFSCar)

 Oscar Peitl (1996)

Mastelaro VR and Zanotto ED. Residual stresses in a soda-lime-silica glass-ceramic. *Journal of Non-Crystalline Solids*. 1996; 194(3):297–304. http://dx.doi.org/10.1016/0022-3093(95)00509-9.

Peitl O, Zanotto ED and Hench LL. Highly bioactive P_2O_5-Na_2O-CaO-SiO_2 glass-ceramics. *Journal of Non-Crystalline Solids*. 2001; 292(1–3):115–126. http://dx.doi.org/10.1016/S0022-3093(01)00822-5.

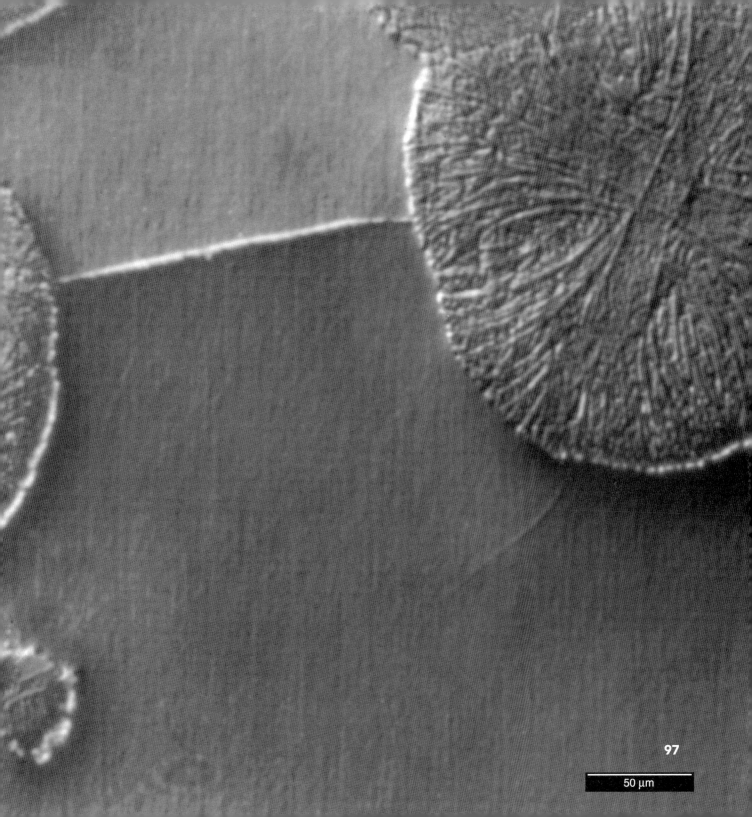

97

50 μm

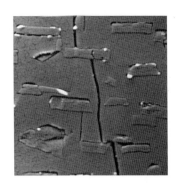

Toughening of a Glass-ceramic by Crack Deflection

One of the main toughening mechanisms afforded by glass crystallization is crack deflection by the embedded crystals. Such effect is clearly shown in this micrograph of micrometer-sized mica-type crystals in a glass-ceramic.

Scanning electron microscopy.

 OSI-FSU (Jena, Germany)

 Courtesy of Christian Ruessel

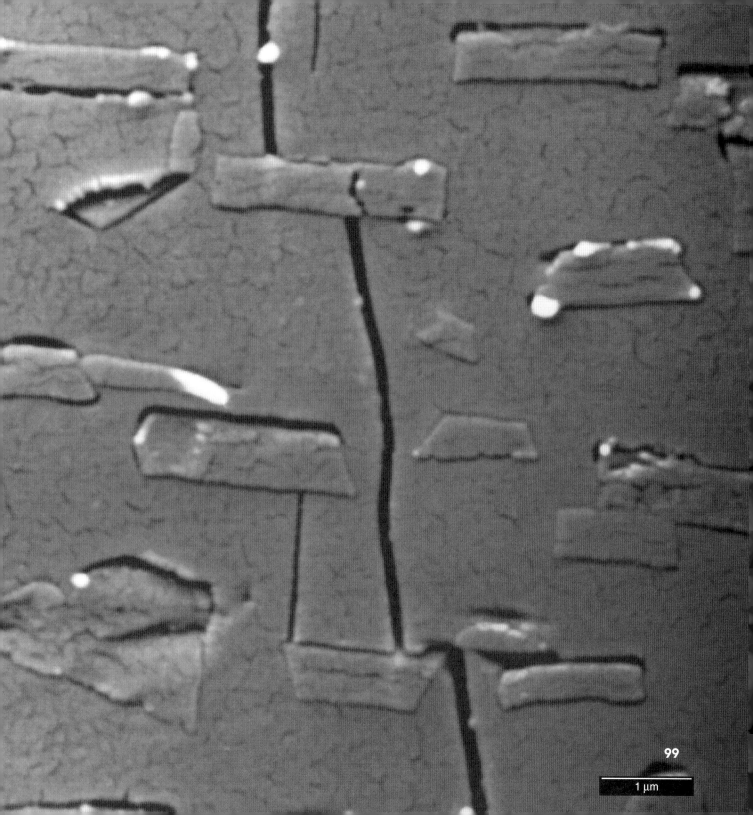

99

1 µm

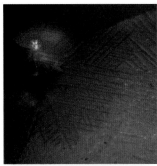

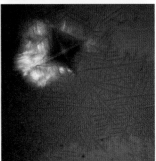

Toughening of a Dental Glass-ceramic by Crack Deflection

One of the main toughening mechanisms of brittle materials is the deflection of cracks around the crystals. This effect is illustrated in this micrograph of crystals of 20 μm in a lithium disilicate glass-ceramic intended for dental use. The material was indented with a pyramidal tip, and the cracks thus generated were diverted from their natural course by the intricate crystals. The fracture toughness of glass is typically between 0.50 and 0.75 MPa.m$^{1/2}$, while that of this glass-ceramic is about 2.8 MPa.m$^{1/2}$.

Optical microscopy, reflected light.

 LaMaV (UFSCar)

 Mari Villas-Boas and Rafa Carioli (2011)

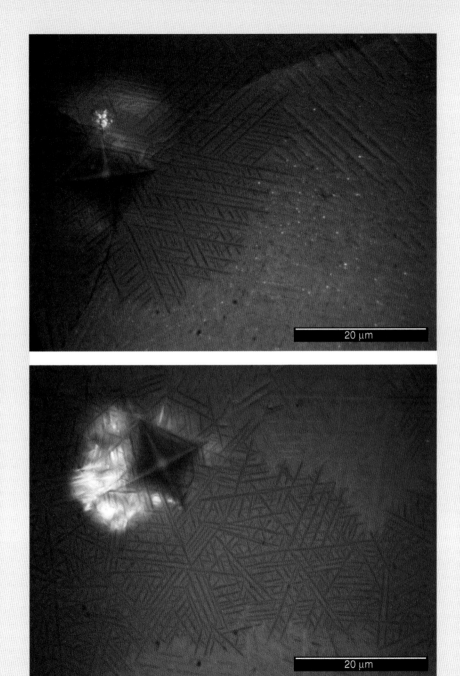

20 μm

20 μm

Cracks and Bubbles in Glass-ceramics

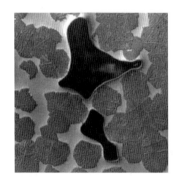

Nucleation of Bubbles in a Bio Glass-ceramic

The crystallization of glasses often gives rise to (undesirable) bubbles. This is because certain gases (e.g., O_2, N_2, H_2O, and CO_2) are dissolved usually in small amounts in glass. During a heat treatment above T_g, when the crystallized fraction substantially increases, often bubbles are nucleated and grown in the glass matrix because residual gases are less soluble in the crystals. In this case, two large bubbles are generated in a bioactive glass-ceramic. The companies keep secret about their strategies to avoid this problem in commercial glass-ceramics.

Optical microscopy.

 University of Florida (EUA) and LaMaV (UFSCar)

 Oscar Peitl (2000)

Peitl O, Zanotto ED and Hench LL. Highly bioactive P_2O_5-Na_2O-CaO-SiO_2 glass-ceramics. *Journal of Non-Crystalline Solids*. 2001; 292(1–3):115–126. http://dx.doi.org/10.1016/S0022-3093(01)00822-5.

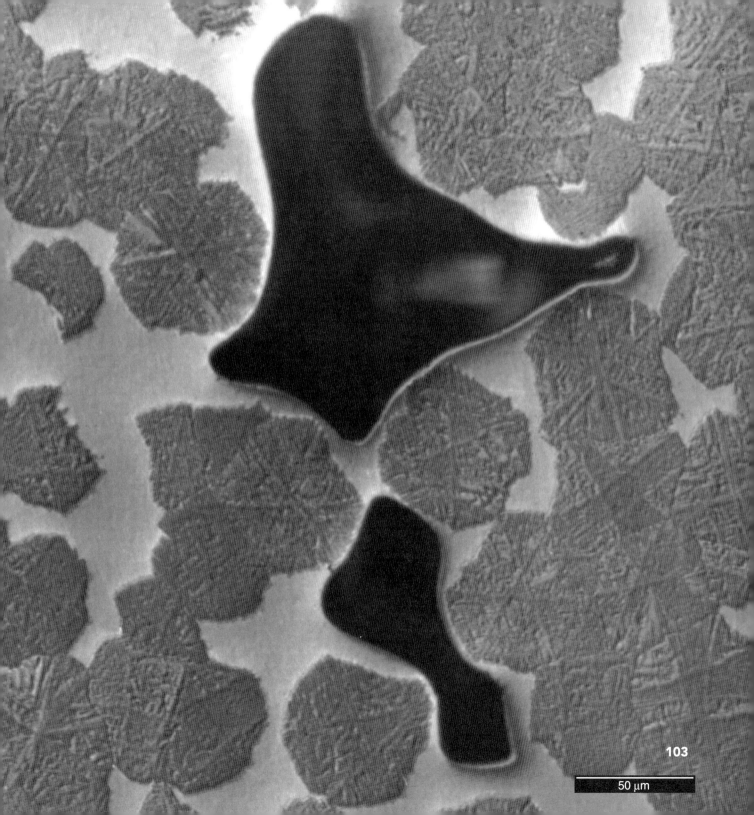

103

50 μm

REVIEWS

of "Crystals in Glass: A Hidden Beauty"

"The book is primarily a collection of selected but representative exemplary micrographs of modern glass-ceramic materials, but into an easy-to-read background of science and historical account. As such, it provides convenient tea-table entertainment for materials scientists, chemists and physicists, but also a motivating and captivating introduction into materials science in general, and glass-ceramics in particular. Its careful balance of illustrations and text should be kept as it is.

The author is one of the internationally leading experts in the field of glass-ceramic materials.

The book (its Portuguese version) is well-written (as far as one can say this, since the main purpose is its illustrations). It appears carefully structured and well-balanced between limited amounts of text while keeping the necessary scientific standard."

Anonymous reviewer

Crystals in Glass: A Hidden Beauty, First Edition. Edgar D. Zanotto.
©2013 John Wiley & Sons, Inc. and The American Ceramic Society.
Published 2013 by John Wiley & Sons, Inc.

"Let me start by stating that there is absolutely no doubt that Prof. Zanotto is an international expert on the topic of crystallization in glasses, and that he is well qualified to write a book — any book — on this topic. He has a long career dedicated to this type of research, and has gathered valuable data and understanding on a topic of critical importance for the pure and applied glass communities.

Let me also state that I have no doubt that the book, in its Portuguese version, is well done. I looked at some of the online images, and found them to be quite beautiful. The amount of work in selecting the images was clearly substantial, and it does not take into account the years of research that led to the images in the first place. As a "coffee table book", I see great appeal for the subset of photographers and art enthusiasts that focus on science.

Where I do have a concern, however, is in the overall thrust and audience of the book. Prof. Zanotto states that the book "...*covers very significant discoveries about the phenomenon of crystallization of glasses in the past 35 years...*", which coincides with the length of his professional career, as mentioned later in the proposal. The primary audience is said to be "academic researchers, professors, science students, engineers, geologists, chemists, physicists", while the secondary audience is "…photographers, artists, and the general public interested in arts…". But the book, as it stands now, contains no equations, tables, graphs, or molecular structures. This puzzles me, as it seems to miss materials that the primary audience of scientists would find interesting.

I take this to mean that the book really is a "coffee table book", and that the primary audience is such mainly because scientists will like the beautiful images and relate better to them. I also believe that the book is based exclusively on the work of Prof. Zanotto, which is extensive, but does not encompass the entire field.

Under the understanding above, I think the book would appeal to a (fairly limited) primary audience, and perhaps an even smaller section of the general public. The very beautiful images will indeed provide an attractive sale "hook", though scientists may find the content less satisfying.

In general, I find the proposal good. The book ought to be published, though after some thought and discussion on the precise audience. In particular, the depth of the scientific content and the breadth of material (limited to the author's work) ought to be considered. Nevertheless, it is my hope that an ultimate book goes forward—I would certainly consider owning a personal copy."

Anonymous reviewer

"The book shows special phenomena of crystal formation in glasses. This is based on fundamentals of nucleation and crystallization mechanisms in glass. The author shows the phenomena of these mechanisms. It is not the interest of the author to explain the phenomena. Therefore, the book is not a textbook, but shows very interesting phenomena. This idea of the book will encourage students and scientists to study the phenomena of glass-crystallization in more detail.

Based on this idea of the author, the book is organized logically. Therefore, I agree with the selection, order of presentation, and weighting of topics.

The idea of the author, Prof. Dr. E.D. Zanotto, is to show the microstructure he investigated with his students and co-workers and in collaboration with scientists in the world. Therefore, other results than these are not demonstrated. But this is O.K. and the author shows the wide area of glass-ceramic research with reporting on activities of a technical committee of the International Commission on Glass (ICG). The names of most important research groups worldwide are shown. One subject was omitted and should be included: reference to existing textbooks on glass-ceramics demonstrated as a short reference list at the end or at the beginning of the book.

Another aspect should also be considered: D. Schulze has published a very interesting book with the title "looking, understanding, designing" (with sub-title: microstructures investigated with the electron microscope). This book is published in German language: "Sehen, Verstehen, Gestalten" Werkstoff-Informationsgesellschaft mbH, 1998. A comment to this book would be helpful to the reader. Also, Prof. Zanotto should comment on the book "Glasfehler" written by W. Vogel and published by Springer in 1993.

Based on the idea, to present results of the group of Prof. Zanotto, no subjects or topics should be deleted.

The phenomena are explained correctly, that is, the content is technically correct, and current. Some figures do not contain a bar showing the magnifications. This should be changed. Also, it would be very helpful for the reader to add into the caption of the figures: the method of preparation (polished, fracture surface, etched etc.) and the used method (SEM, optical microscopy etc.)

Prof. Zanotto is well-known in the area of fundamentals of glass crystallization. Here, I know him personally and respect very much his deep fundamental understanding of nucleation and crystallization theory.

All in all, I would recommend publishing the book. Please consider some aspects I mentioned and ask Prof. Zanotto to carry out these minor revisions of the text."

Anonymous reviewer

"I enjoyed reviewing the proposal for Prof. Zanotto's book "Crystals in Glass: A hidden beauty". The book certainly presents a unique and, literally, quite attractive perspective on the interplay of the glass transition and crystallization.

I am not aware of a similar publication elsewhere, nor did I appreciate before how visually attractive crystallization of vitreous samples could be. I think that, similarly, many specialists and amateurs alike will appreciate both the aesthetic and scientific aspect of the book. It helps that Prof. Zanotto is a well-known and respected figure in the field.

The book is also interesting in that it contains data from a variety of physico-chemical situations, in terms of distinct morphologies of the crystals, crystal-nucleation in bulk, surface, etc., and several chemically distinct systems. It is highly commendable that the author provides citations to the original articles where these data were published. In addition to adding credibility to the publication, this also serves to popularize the field to researchers from other field and young scientists.

I should like to point out that the book should also be marketed to members of the American Chemical Society (ACS) and American Physical Society, in addition to the potential audience mentioned in the proposal. Note the ACS is the largest professional society in the world. I think that pharmaceutical researches will also enjoy the book.

To summarize my review, I myself would be interested in acquiring the book and believe many others would be also."

Anonymous reviewer

ABOUT THE AUTHOR

Edgar Dutra Zanotto

Edgar Dutra Zanotto (EDZ) is a Professor of Materials Science and Engineering, head of the Vitreous Materials Lab and director of the Center for Research, Technology and Education in Vitreous Materials (CeRTEV) of the Federal University of São Carlos, Brazil.

The fundamental research activities of EDZ and his collaborators comprise the theme "*crystallization kinetics and properties of glasses and glass-ceramics.*" Their research work encompasses development of new glasses, rigorous tests, improvement or development of nucleation and growth models for glasses, the effects of liquid phase separation on nucleation, surface crystallization kinetics, overall crystallization, glass stability *versus* glass-forming ability, the correlation between molecular structure and nucleation mechanism, sintering with simultaneous crystallization, diffusion processes controlling crystallization, and mechanical, rheological, and thermal properties of glasses and glass-ceramics.

The technological research of Zanotto's team includes projects in partnership with 22 companies over the past two decades. He is most interested in the development of new glasses and glass-ceramics with improved or new functionalities. He has 12 registered patents, two of which have been awarded prizes by IBM, and at the Invention 1996 "State Governor Prize."

EDZ has published about 250 original and review papers in periodicals and conference proceedings, 20 book chapters and 2 books, and advised 65 MSc, PhD, and postdoctoral research projects plus about 80 scientific initiation projects. Regarding his main research focus (phase transformations in oxide glasses), the Scopus database indicates that he ranks first with the keywords "crystal nucleation growth glass" among the world's most prolific.

EDZ is an Editor of the *Journal of Non-Crystalline Solids* and member of the international advisory boards of four other periodicals: *International Journal of Applied Glass Science*, *Materials Research*, *Buletin de la Sociedad Espanhola de Ceramica y Vidrio*, and *Cerâmica*. He is a member of the World Academy of Ceramics, the Science Academy of the Developing World (TWAS), the Brazilian Academy of Sciences, and the São Paulo State Academy of Sciences, and a fellow of the Society of Glass Technology (UK). He has chaired seven important international glass congresses and delivered about 100 invited talks at national and international scientific conferences, which include 15 plenary speeches, plus about 50 invited seminars at universities and industries.

He received 25 awards, including the Brazilian Order of Scientific Merit (*Comendador = knight*) in 2006 from the Brazilian President Luis Inacio Lula da Silva, the TWAS Engineering Prize 2011, the Admiral Álvaro Alberto Prize 2012 from CNPq, FCW and the Brazilian Navy (the most prestigious award for R & D in Brazil), and three of the most important international glass research awards (the *Zachariasen Award* 1990 from the *Journal of Non-Crystalline Solids*, the *Vittorio Gottardi Memorial Prize 1993* from the *International Commission on Glass*, and the *George W. Morey Award 2012* from *The American Ceramic Society*).

His contributions to the academia, industry, and society have been exerted by several executive and consulting positions: former adjunct coordinator of exact sciences and engineering at the São Paulo State Research Foundation (Fapesp); manager of Fapesp's patent and licensing office; co-founder of the Brazilian Materials Research Society (SBPMat) and of the scientific journal Materials Research (editor-in-chief during 13 years).

Currently he is a curator of the São Carlos City High Technology Park; counselor emeritus, past vice president, and currently director of the Brazilian Ceramic Society; council member of the São Paulo Academy of Sciences; founder and president of Vitrovita-the glass-ceramic innovation institute; chairman of the TC7 (Crystallization Committee of the International Commission on Glass (ICG)); Brazilian representative to the International Ceramic Federation (ICF), and member of the Advisory Board of the International Materials Institute (NSF - USA).

Edgar Dutra Zanotto
Vitreous Materials Lab
Department of Materials Engineering
Federal University of São Carlos
lamav.weebly.com

São Carlos, SP, Brazil
May 2013